Intelligent Buildings and Building Automation

Intelligent Buildings and Building Automation

Shengwei Wang

Spon Press
an imprint of Taylor & Francis

LONDON AND NEW YORK

First published 2010
by Spon Press
2 Park Square, Milton Park, Abingdon, Oxon OX14 4RN

Simultaneously published in the USA and Canada
by Spon Press
270 Madison Avenue, New York, NY 10016, USA

Spon Press is an imprint of the Taylor & Francis Group, an informa business

Typeset in Sabon by
Pindar NZ, Auckland, New Zealand

This publication presents material of a broad scope and applicability. Despite stringent efforts by all concerned in the publishing process, some typographical or editorial errors may occur, and readers are encouraged to bring these to our attention where they represent errors of substance. The publisher and author disclaim any liability, in whole or in part, arising from information contained in this publication. The reader is urged to consult with an appropriate licensed professional prior to taking any action or making any interpretation that is within the realm of a licensed professional practice.

British Library Cataloguing in Publication Data
A catalogue record for this book is available from the British Library

Library of Congress Cataloging-in-Publication Data
Wang, Shengwei.
Intelligent buildings and building automation / Shengwei Wang.
    p. cm.
    Includes bibliographical references and index.
    1. Intelligent buildings. 2. Buildings—Mechanical equipment—Automatic control. I. Title.
    TH6012.W36 2010
    696—dc22   2009018512

ISBN10: 0-415-47570-8 (hbk)
ISBN10: 0-415-47571-6 (pbk)
ISBN10: 0-203-89081-7 (ebk)

ISBN13: 978-0-415-47570-9 (hbk)
ISBN13: 978-0-415-47571-6 (pbk)
ISBN13: 978-0-203-89081-3 (ebk)

Contents

Figures and tables

Figures

Tables

Preface

Intelligent building (IB) and building automation (BA) systems play an essential role in most sophisticated modern buildings. Monitoring and automatic control of building services systems are important to ensure that the design objectives are met in operation. Graduates and engineers associated with building systems need an adequate knowledge and understanding of IB and BA systems, the associated technologies and their features, as well as their implementation. Based on my experience of teaching this subject area to building services engineering students and facilities management students over the last 16 years, I feel that there is a lack of a comprehensive reference book addressing this need from the viewpoint of building services or heating, ventilating and air-conditioning (HVAC) engineers and facilities engineers and managers.

This book provides readers with an explanation of the state of the art in IB/BA systems and technologies, and enables them to understand the working principles and applications of BA systems and the control of building services systems. It mainly addresses the following issues.

- Progress and state of the art in IB/BA systems, and their configuration and integration.
- BA network, including wired/wireless local area networks (LAN) and Internet, communication protocols and standards as well as their applications.
- The interfacing and integration of BA subsystems with building services systems.
- Process control and tuning of local control loops.
- The control and optimization as well as the operational characteristics of typical HVAC systems, including air-conditioning systems and central chilling systems.
- The automation systems for lighting-system control, security and access control, and fire safety control.

It is hoped that this book will provide an effective reference text for engineers and students in building services engineering (architectural engineering, building environmental engineering) to enable them to understand the technology and implementation of building automation systems, as well as to help engineers and students working on IB/BA systems and technology to understand the operation and control of major building services systems. My aim has been not to present a handbook listing all of the systems and technologies, but to provide a reference book giving the reader a clear picture and understanding of IB/BA systems, the commonly used technologies and the main issues concerning their applications. Readability and effectiveness in supporting the readers' learning have been among the major concerns in organizing and selecting the material for this book.

Acknowledgements

I would like to acknowledge the assistance of my PhD students and post-doctoral fellows in drafting the manuscript and drawings, in particular Dr Zhengyuan Xu on the chapters associated with network standards, Professor Dr Xinhua Xu, Dr Zhenjun Ma, Dr Qiang Zhou and Dr Gongsheng Huang on various topics, the support of my colleague, Mr Daniel Wah-tong To, in reviewing the chapter on lighting-control systems, and Dr Linda Fu Xiao in reviewing the chapter on security and safety control systems, as well as the advice and suggestion of my friend, Professor Arthur Dexter, Department of Engineering Science, Oxford University, on the overall organization of this book. This book is to a large extent based on the teaching materials used for two subjects, Engineering Intelligent Buildings, and Building Automation and Control, over the last few years in The Hong Kong Polytechnic University. I have found feedback from students to be particularly useful in the writing of this book.

Professor Dr Shengwei Wang
Chair Professor of Building Services Engineering
Department of Building Services Engineering
The Hong Kong Polytechnic University
Kowloon, Hong Kong
beswwang@polyu.edu.hk

1 Introduction to intelligent buildings

1.1 Definitions of intelligent building

The concept of intelligent building (IB) has received increasing attention over the last two decades, as various intelligent buildings and IB technologies have been developed and people have come to understand IBs. Many definitions have been suggested during this period, but as the building industry and information technology develop, what an IB contains is changing too.

It is difficult to formulate a unique conception of IBs and no single definition is accepted worldwide. However, it is not necessarily important to have a standard definition of IB, although it is vital to have a clear understanding of what different people are talking about when this terminology is used. Different countries and regions and different disciplines may have diverse preferences and different IB concepts may predominate. However, the approaches to defining an IB can be grouped into three categories as listed below:

1   performance-based definitions;
2   services-based definitions;
3   system-based definitions.

Some definitions representative of these categories are discussed in the following sections, which will help readers gain a general understanding of IBs.

1.1.1 Performance-based definitions

Performance-based definitions define IBs by stating what performances a building should have. A typical performance-based IB definition may be that of the European Intelligent Building Group (EIBG). EIBG (located in the United Kingdom) defines an IB as a building created to give its users the most efficient environment; at the same time, the building utilizes and manages resources efficiently and minimizes the life costs of hardware and facilities.

Another example of a performance-based definition is that given by the Intelligent Building Institute (IBI) in the United States, which states that an

IB provides a highly efficient, comfortable and convenient environment by satisfying four fundamental demands: structure, system, service and management, and optimizing their interrelationship.

Performance-based IB definitions emphasize building performance and the demands of users rather than the technologies or systems provided. According to this category of definition, owners and developers of buildings need to understand correctly what kind of buildings they want and also how to satisfy continuously the increasing demands of users. Energy and environmental performances of buildings are certainly among the important issues of an IB. An intelligent building should also adapt itself quickly in response to internal and external conditions, and to meet the changing demands of users.

1.1.2 Services-based definitions

Services-based definitions describe IBs from the viewpoint of services and/or quality of services provided by buildings. The Japanese Intelligent Building Institute (JIBI) provides an example of a services-based definition: an IB is a building with the service functions of communication, office automation and building automation, and is convenient for intelligent activities. Services to users are emphasized. The key issues of IBs in Japan focus on the following four services aspects:

1   serving as a locus for receiving and transmitting information and supporting efficient management;
2   ensuring satisfaction and convenience of persons working inside;
3   rationalization of building management to provide more attractive administrative services at lower cost;
4   fast, flexible and economical responses to the changing sociological environment, diverse and complex working demands and active business strategies.

1.1.3 System-based definitions

System-based IB definitions describe IBs by directly addressing the technologies and technology systems that IBs should include. A typical system-based IB definition is the one suggested in the Chinese IB Design Standard (GB/T50314–2000), which states that IBs provide building automation, office automation and communication network systems, and an optimal composition integrates the structure, system, service and management, providing the building with high efficiency, comfort, convenience and safety to users.

A more straightforward system-based IB definition has been used by some professionals and developers in practice. It labels the IBs as '3A', which represents building automation (BA), communication automation (CA) and office automation (OA).

1.1.4 How to make a building intelligent in reality

With so many different definitions of, and views on, intelligent buildings, it is difficult to suggest a unique and definitive description of IBs. It is also not particularly necessary. Readers do not need to worry about what IB definition we should have; rather, there is one important question we should ask ourselves: how do we make a building intelligent in reality? This is a definite goal of IBs, and trying to answer the question will help us to have a better understanding of the contents of IBs.

Readers may appreciate that buildings which can be considered as intelligent or smart might not necessarily have technology systems, as there have been buildings constructed even long ago that provided rather smart functions. Readers may also agree that a building fully equipped with technology systems might not be intelligent in reality if the systems cannot be coordinated or they do not function properly.

However, in the context of the modern building environment, it is obvious that intelligent buildings cannot exist without involving technology systems, especially information technology (IT) systems. But having those technology systems is not enough to make a building an intelligent one. Furthermore, the technology systems should be correctly configured and properly integrated with each other and with the building facilities. The system functions should be appropriately customized to meet user requirements and to provide the expected performance of intelligent buildings.

Finally, the technology systems, including their integration and interoperation, should be properly commissioned and maintained to ensure they function as expected. Besides the system hardware and software, the application software, including that for facility automation and control, optimization and management, should be customized and commissioned appropriately. A building may have technology systems, but if they are not working correctly it will not make the building intelligent in reality. Instead, the technology systems may create headaches for operators and users.

IBs are interdisciplinary and involve multi-industrial system engineering. They require the right combination of architecture, structure, environment, building services, information technology, automation and facility management. In addition, IBs are also strongly related to economic and cultural aspects.

The definitions and concepts discussed in this section are mainly from the viewpoint of building facility systems. In fact, professionals from different building sectors also have different views on the concept and contents of intelligent buildings. In the following section, some views of architects and structural engineers are discussed.

1.2 Intelligent architecture and structure

Although the successful use of advanced technologies, including IT, is the main feature of intelligent buildings, the implementation of technologies should not be the sole objective of IBs. Performance is definitely a key objective of intelligent buildings, although performance can be interpreted very differently as discussed above. As regards the hardware facilities, intelligent buildings cannot be separated from the architecture design, building façades and materials, which are among the essential elements of intelligent buildings.

1.2.1 Intelligent architecture

Intelligent architecture refers to built forms whose integrated systems are capable of anticipating and responding to phenomena, whether internal or external, that affect the performance of the building and its occupants. Intelligent architecture relates to three distinct areas of concern:

1   intelligent design;
2   appropriate use of intelligent technology;
3   intelligent use and maintenance of buildings.

Intelligent design requires that the building design responds to humanistic, cultural and contextual issues; that it exhibits simultaneous concern for economic, political and global issues; and that it produces an artificial enclosure which exists in harmony with nature. Existing in harmony with nature includes responding to the physical laws of nature and the proper use of natural resources.

Appropriate use of intelligent technology. The mere availability of a large variety of smart materials and intelligent technologies often results in their use in inappropriate situations. Integrating intelligent technologies with an intelligent built form that responds to the inherent cultural preferences of the occupants is a central theme in intelligent architecture. As an example, in areas where people place a high premium on operable windows for conservation of electricity, the most appropriate and efficient air-conditioning strategy for a building may be the use of thermal mass and night-time free cooling instead of a high-tech air-conditioning system. In other cases, the use of carefully selected electric lighting and environmental control strategies may be more appropriate.

Intelligent use and maintenance of buildings. Truly intelligent architecture incorporates intelligent facility management (FM) processes. For a design to be intelligent it must take into consideration the life cycle of a building and its various systems and components. Although an intelligent building may be complex, it should be fundamentally simple to operate, be energy and resource efficient, and easy to maintain, upgrade, modify and recycle.

Materials and equipment that require complex maintenance and unhealthy cleaning agents, and building components that must be treated as hazardous waste in the recycling process (e.g. mercury in light-bulbs) would not be used in a fully developed intelligent architecture.

1.2.2 Intelligent and responsive building façades

The character of the building envelope will be affected dramatically by the development of intelligent buildings. Façades designed to integrate a host of emerging technologies will have an inherent 'intelligence' and be able to respond automatically, or through human intervention, to contextual conditions and individual needs. Intelligent façades currently can:

- be centrally controlled while still providing the occupant with the ability to manually override the system;
- change their thermophysical properties such as thermal resistance, transmittance, absorptance, permeability, etc;
- modify their interior and exterior colour and/or texture;
- function as communicating media façades with video and voice capabilities;
- change optical properties and allow the creation of patterned glazing, providing the opportunity for dynamic shading and remote light control.

The development of the intelligent and responsive façade necessitates the redefinition of the terms 'window' and 'wall'. With the introduction of new glazing and wall assemblies, what is 'transparent' may become 'opaque' with the flick of a switch. Central controls for intelligent façades will respond to climatic conditions by transforming the building envelope to optimize heating and cooling loads, daylight utilization, natural ventilation, and so on. Intelligent façades will transport daylight deep into a building's interior and allow the occupants to determine the degree of luminous, acoustical and thermal comfort required along with the degree of visual and acoustical privacy provided by the enclosure. Additionally, we can now imagine interior partitions that will allow the occupants to transform the aesthetic quality of their working environment whenever and however they choose.

The idea of the intelligent or smart system, originally applied to electrical, mechanical and aerospace systems, recently has been extended to include civil structures as advances in sensing, networking and new materials have made continuous monitoring and control of structural functions a realizable goal. By definition, the intelligent structure has the capability to identify its status and optimally adapt its function in response to stimuli. The major focus of the intelligent civil structure has been on two areas:

1 identification of structural behaviour or properties (e.g. deformation, energy usage or damage evaluation);

2   control of structural response to stimuli, whether external (e.g. wind or earthquake) or internal (e.g. acoustics or temperature variation).

1.3 Facilities management vs. intelligent buildings

The usual definition of facility management, commonly abbreviated as FM, is the practice of coordinating the physical workplace with the people and work of the organization; it integrates the principles of business administration, architecture and the behavioural and engineering sciences. The definition is often simplified to mean that facility managers integrate the people of an organization with its purpose (work) and place (facilities).

The International Council for Research and Innovation in Building and Construction (CIB) Working Commission on Facilities Management and Maintenance summarized the scope of facilities management in the following categories:

- *Financial management.* This refers to the investment issues including: sale and purchase, rental return, rebuild or renovation, etc.
- *Space management.* This includes space utilization, interior design, fit-out and relocation, etc.
- *Operational management.* This refers to the maintenance management and refurbishment and lease and property management including building enclosure, building services, building environment and building grounds.
- *Behavioural management.* This refers to the users of the building, including users' perceptions, the satisfaction of the occupants and participation of users, etc.

Facility management is also often referred to as a profession or professional discipline, which has received more and more recognition over the last two or three decades. It is, in fact, a fairly new business and management discipline.

Intelligent building and facilities management are closely linked. The scope of facilities management defined by FM professionals often includes significant parts of IB hardware facilities and functions. On the other hand, the contents of intelligent buildings defined by IB professionals often include significant FM elements. This situation reflects the fact that definitions of both terminologies cover a very wide scope and different points of view. In fact, modern IB systems are complex and powerful systems offering various functions for building control and management. The IB system is a preferred platform for supporting various tasks of building facilities management. At the same time, the success of implementing FM functions in IB systems makes intelligent building more attractive. IB systems as complex facilities to be managed actually create business opportunities for FM.

1.4 Technology systems and evolution of intelligent buildings

The evolution of intelligent building systems is illustrated in Figure 1.1, which is modified and updated from the 'Intelligent Building Pyramids' developed by the European Intelligent Building Group. The pyramid illustrates the contents and evolution of IB technology over the last few decades. The pyramid is open at the top, emphasizing that the intelligent building systems are not enclosed within buildings any more but instead are merged with IB systems in other buildings as well as other information systems via the global Internet infrastructure.

Intelligent buildings began from the automatic intelligent control of typical building services processes and communication devices. Along with the rapid evolution of electronic technology, computer technology and information technology, intelligent building systems are becoming more and more advanced, and the level of integration is being developed progressively from the subsystem level to total building integration and convergence of information systems.

Before 1980, the automation of building systems was achieved at the level of the individual apparatus or device. After 1980, intelligent building systems entered the integrated stages. There has been great progress on IB system integration in terms of both technology and scale. IB systems after 1980 can be divided into five stages as follows:

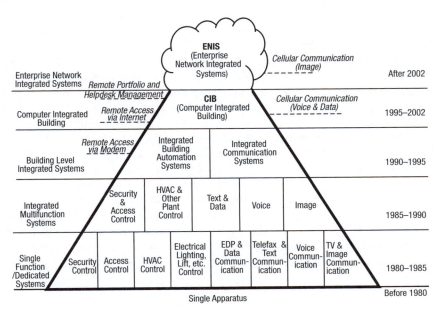

Figure 1.1 Intelligent building pyramid.

1   integrated single function/dedicated systems (1980–5);
2   integrated multifunction systems (1985–90);
3   building level integrated systems (1990–5);
4   computer integrated building (1995–2002);
5   enterprise network integrated systems (2002–).

At the stage of *integrated single function/dedicated systems* (1980–5), all the BA subsystems (including security control; access control; heating, ventilation and air-conditioning [HVAC] control; lighting control; lift control; other electrical systems; fire automation; etc.) and CA subsystems (including electronic data processing [EDP]) and data communication; telefax and text communication; voice communication; TV and image communication; etc.) were integrated at the level of a single or individual function subsystem. Integration and communication between the automation systems of different subsystems was impossible.

At the stage of *integrated multifunction systems* (1985–90), security and access control were integrated. The automation systems of building plants or services systems were integrated. There were unified networks for text and data communication, voice communication and image communication respectively. At this stage, the integration of systems with the same nature or similar functions was achieved.

At the stage of *building level integrated systems* (1990–5), both BA and communication systems were integrated at building level as building automation system (BAS) and integrated communication system (ICS). At this stage, a BA system could be accessed remotely via telephone network using a modem, while the cellular phone for voice and data communication was introduced to the market.

At and after the stage of *computer integrated building* (1995–2002), convergence networks became available and were used in practice progressively, thanks to the popular use of Internet protocol (IP) network technologies and increased network capacity. At this stage, the integration was at the building level. Remote monitoring and control could be achieved via the Internet.

At the stage of *enterprise network integrated system* (2002–), the intelligent systems can be integrated and managed at enterprise level or city level. Intelligent building systems are not enclosed within buildings any more; they are merged with IB systems in other buildings as well as other information systems via the global Internet infrastructure. Integration and management at this level become possible due to the applications of advanced IT technologies such as Web Services, XML, remote portfolio management and helpdesk management, among others. In terms of communication, image communication via cellular phone has been brought into practical use.

1.5 Concluding remarks on IB systems

The integration of IB components and subsystems has been the trend of IB technology development. Integration is essential for most functions of IB systems, such as automatic monitoring and management, and building performance optimization and diagnosis. Function integration increases the flexibility and possibilities of intelligent management of buildings. The integration of the automation and control systems is the basis for function integration. Digital technology plays a very important role in the integration as systems that consist of traditional technologies have many constraints in terms of information exchange and integration. The microprocessor, providing amazing power in computation, and in transmitting and processing information, is the key element of digital systems and the key element of IB and BA systems.

Modern IB systems have been becoming very large in terms of system scale and complex in terms of hardware and software system configurations, while their functions and capacities have been increasing progressively. System reliability is an important issue. Utilizing a decentralized network or a decentralized local area network (LAN) is the key to solving the system reliability issue and simplifying IB networks. Distributed intelligence is a major philosophical solution to ensure the reliability of such complex IB and BA systems. 'Integrated but independent' is one of the most essential concerns in the development and configuration of IB and BA systems.

References

DEGW and Tekinibank. (1995) *The Intelligent Building in Europe*, London and Milan: British Council of Offices, The College of Estate Management.

Himanen, M. (2003) *The Intelligence of Intelligent Buildings*, Finland: VTT Publications.

Kroner, W. M. (1997) 'An intelligent and responsive architecture', *Automation in Construction*, 6: 381–93.

So, A. T. B. and Chan, W. L. (1999) *Intelligent Building Systems*, Boston: Kluwer Academic Publishers.

Wong, J. K. W., Li, H. and Wang, S. W. (2005) 'Intelligent building research: a review', *Automation in Construction*, 14(1): 143–59.

2 Digital controllers

Building automation systems (BAS), also known as building management systems (BMS), are principally integrated processor-based systems. A BAS outstation is actually a digital controller that is linked into the entire BAS via a network. In principle, a BAS outstation (or digital controller) is a micro-computer system specially designed to be suitable for data acquisition, control and communication and other functions.

Besides the digital controller, there are other four typical types of controller, including mechanical controller, pneumatic controller, electrical controller and electronic controller. They are still used nowadays in buildings but are less popular than the digital controller.

This chapter presents the basic principles of processors, the structure of digital controllers and data acquisition, and an introduction to sensors and actuators.

2.1 Data form used in computers

People are now used to handling data in decimal form. Over the centuries, different forms have been used to deal with data by different people for varied purposes. For instance, the hexadecimal form has been used in China for thousands of years and it is still used in street markets in Hong Kong today. Digital computers operate exclusively on data in binary form due to its ability to be handled easily by electronic circuits. That means the machine only needs to recognise two states: *on/off* or *high/low*. These binary states are usually designated as *0* and *1* for recording purposes. Different voltage is often used to code these two states. It is a reliable way for digital computers to handle information, and it is particularly suited to digital electronic circuit design.

Figure 2.1 shows the *0* and *1* coded as voltage pulses. In the case of a *1*, the line voltage in the communication cable is raised as rapidly as possible from zero to a certain steady-state value and remains constant at that value for a specific time. A *0* is represented by lack of such a pulse (zero volts). A minimum period is necessary to provide sufficient steady-state voltage for reliable data transmission, which is strongly related to the performance of

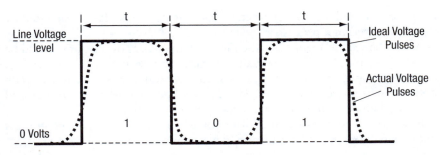

Figure 2.1 Representation of binary data by voltage pulses.

Table 2.1 Comparison of data in different forms

Binary	Decimal	Hexadecimal
000000	00	00
000001	01	01
000010	02	02
000011	03	03
000100	04	04
000101	05	05
000110	06	06
000111	07	07
001000	08	08
001001	09	09
001010	10	0A
001011	11	0B
001100	12	0C
001101	13	0D
001110	14	0E
001111	15	0F
010000	16	10
010001	17	11
010010	18	12
010011	19	13
010100	20	14

the processor and to cable characteristics. The period is often designated by the clock speed of a processor.

As can be seen in Figure 2.1, the ideal voltage pulses of the data should be the sharp step pulses. However, due to cable and processor characteristics, the actual voltage pulses which appear in the communication cable deviate noticeably from the ideal pulses. The data (voltage pulses) can be recognized correctly before such deviation reaches a certain degree.

It can be considered that all information is represented in digital computers in this way (note that data transmitted between computers can be in different formats). The information refers to *data*, *instructions* and *addresses*, although they have very different uses. The actual *data*, *instructions* and *addresses* are represented by groups of 0s and 1s. Table 2.1 shows a comparison of some data in decimal, binary and hexadecimal forms.

2.2 Microcomputer

2.2.1 The microprocessor

The microprocessor or central processing unit (CPU) is the principal component both of the conventional microcomputer and the BAS control station. It is a micro-electronic chip produced by large-scale integrated (LSI) circuit manufacturing techniques on a small chip of silicon (about 5 cm²), and it is the 'brains' of a microcomputer. More recent microprocessors are VLSI (very large scale integrated) chips and are correspondingly more powerful. The actual microprocessor chip is often contained in a package with a number of pins, like legs, connecting the chip to the motherboard.

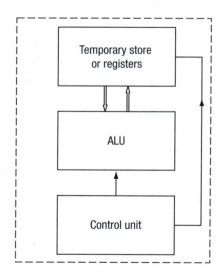

Figure 2.2 Central processing unit (CPU) or microprocessor.

The LSI or VLSI chips actually contain a large integrated logic circuit based on a large number of resistors, transistors and the like. A complete processor is nowadays made within a single VLSI chip, while, in the early stages, a CPU would have comprised a number of circuit boards.

A simplified schematic diagram of a microprocessor is shown in Figure 2.2. The microprocessor has several basic components: the control unit, a small temporary memory, consisting of registers, and the arithmetic logic unit (ALU).

The *arithmetic logic unit* is the operational unit. It performs calculations, such as addition and multiplication, as well as logical decision-making processes such as selecting and comparing data. The *control unit* controls the operations of the microprocessor and other associated chips, such as memory chips. This unit coordinates all the functions of a microcomputer, and interprets the instructions in a program to perform the control functions necessary to run that program. The *temporary store* is a small memory, composed of a number of registers, which will hold both the data and the immediately required program instructions for the ALU and control unit to work on.

2.2.2 The microcomputer structure and buses

Figure 2.3 illustrates the main essential components and the interconnections of the microcomputer and outstation. The central part consists of three

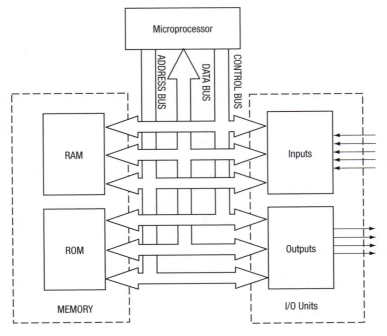

Figure 2.3 Microcomputer principal architecture.

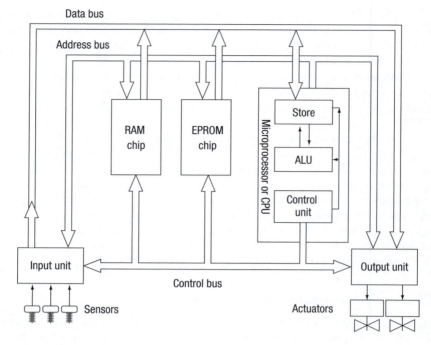

Figure 2.4 Architecture of an outstation.

components: the microprocessor (CPU), the memory, and input and output (I/O) units. All the operations (e.g. data, instructions and address signal transmission) are performed via three buses: data bus, address bus and control bus (bus is an abbreviation of busbar).

Figure 2.4 illustrates how the microprocessor chip of a control station is connected to the memory unit and the input and output units. In fact, all units have their own microelectronic chips. Connection between the microprocessor and the other chips is via the three *buses*. Each of the buses is a set of parallel wires, typically printed on the motherboard.

The data bus is for the transfer of data between chips; e.g. transferring the sensor temperature from the input unit to the memory. The address bus is for locating where the memory or register of the required data is, or where a program instruction is located. The address appears like a telephone number or IP address. Each unit of data stored in the memory unit and all the devices connected to the buses must have an address.

To send data or instruction from an address (A) to another address (B), the following steps will take place.

1 Address A is located by the microprocessor on the address bus.
2 The microprocessor control unit sends a signal via the control bus to take the data from address A to the data bus.

3   Address B is located by the microprocessor on the address bus.
4   The microprocessor control unit sends a signal via the control bus to write
    the data to address B.

(Note that these steps take place at extremely high speed within the
computer.)

As the unit of information or data is a *bit* which corresponds to one-eighth
of a letter, a large amount of binary data is required to represent quite a small
amount of what we perceive as information. Data can be 'packed' into *data
bytes* or *words*.

Processors transmit information between different memory registers
(or component parts) by manipulating the bits of either a byte or a word
simultaneously, which is called *parallel processing*. Data bytes or words are
transferred within the computer in parallel between memory registers or
devices along multiple cables of the buses as the data highway.

The units in which the data transfer speed is usually specified are:

Baud = 1 bit/second   OR   *kBaud = 1000 bits/second*

2.2.3 Memory

Much of the data and the program instructions are stored in the memory
chips. The microprocessor has only a small temporary store, which operates extremely quickly. Therefore larger and consequently slower operating
storage, in the form of separate memory chips, is used. There are two types
of memory chips: read only memory (ROM) and random access memory
(RAM). For microprocessors the 16-bit address bus allows 2^{16} (65,536 =
64K; 1K = 1,024 = 2^{10}) address locations to be handled. These days, address
buses of computers have a greater number of bits, allowing much larger
memory capacity.

The ROM chip can send only data or instructions. It cannot receive and
store them from other chips or I/O ports in the outstation; this memory unit
can only be 'read from', not 'written to'. The ROM chip therefore contains
the manufacturer's program and data which the user cannot alter. The
program and data are 'burnt' into the ROM during manufacturing. As microprocessors can deal only with binary signals, i.e., 1s and 0s corresponding
to high and low voltages, such as 5 V and 0.5 V respectively, the memories
simply have to store 1s and 0s.

The data in ROM is permanent after the chips are 'burnt'. Therefore, the
data will not be lost even if the chips lose power. But the data in RAM will
be lost in this case. Therefore, a battery is often used in BAS outstations to
prevent power loss to memory chips in the event of power failure. The essential functions of the outstation are written and stored by the manufacturer in
ROM chips. Some manufacturers also store the 'standard control functions'

for the outstation, such as the time schedule, on/off control and proportional-integral-derivative (PID) function, in the ROM chips, as a 'library'.

Although ROMs once manufactured are unalterable, a number of manufacturers actually use Erasable and reProgrammable ROMs, namely EPROMs, so that alterations, such as improvements, in outstation models can be made using the same memory chips. The EPROMs are erased with low-intensity ultraviolet light and can be programmed with special equipment.

The central station is often a personal computer and so it is very similar to an outstation except for screen, keyboard and printer. But another significant difference is that the central station also has a considerably larger memory not only in terms of RAM and ROM chips, but in disc storage as well, which is much larger. Much more data can be stored on it and a large amount of sophisticated software can be installed and used on it.

Each binary signal, i.e. a *1* or a *0*, or *ON* and *OFF*, is termed as a binary digit, or bit, and a byte is a group of 8 bits which is treated as a unit and stored at a storage location. An 8-bit microprocessor chip works with data and program instructions in 8-bit, or byte, lengths. More commonly, there are 16-bit microprocessor chips, 32-bit and 64-bit chips being used in PCs. Also common now are 16-bit and 32-bit outstations. A machine that can deal with more bits in a unit, or with a longer word length, is more powerful and has more capabilities.

RAM and ROM chips have typically been able to store 1 to 8 kilobytes (Kbytes or KB; B is a common symbol for byte) each (kilo or simply K here means 2^{10} or 1,024). Modern techniques can now allow one single memory chip to store gigabytes of data (1 GB = 1,024 MB, 1 MB = 1,024 KB). To keep down costs, most outstations do not have extensive memories, although they vary significantly between manufacturers. Consequently, there is a limit to the data that can be stored in an outstation, just as there is a limit to the number of inputs and outputs and programming that it can handle.

Once the memory is full, unless the data was downloaded to the central station, the former readings would be overwritten by the later readings. Care must therefore be exercised with data stored in an outstation.

2.3 Input unit

The microprocessor performs in digital form, while the external devices (e.g. sensors and actuators or valves) are usually analogue. Even if some sensors or actuators operate using digital signals, the signals generally cannot match the microprocessor buses directly. Therefore, an interface is usually required for a microcomputer to communicate with external devices. In BAS outstations, the input and output units provide the communication interface with building services systems.

The input unit takes signals from sensors, relays, meters and the like and converts them into relevant digital signals that the microprocessor can 'understand', of the correct voltage. For instance, a temperature sensor is continuously sending back an electrical signal (either a small voltage or current) to the input section of the outstation. This is an analogue or continuous signal which needs to be converted into bits and bytes to form a digital signal for the CPU to be able to process. The input section therefore contains an analogue to digital (A/D) converter for this purpose.

2.3.1 Sampling

One of the most important functions of any building automation system is the collection of continuous measurement data, at regular time intervals from large numbers of individual measurement sensors, and 'binary' state data from detectors such as smoke alarms. Measured data is generated continuously by individual sensors. However, an outstation can only read the measurements at regular intervals, even though the interval can be rather short. A measured variable is reconstructed in the system from the measurement of these samples. If a lot of measurements are to be 'read' by the outstation, each must be sampled at intervals in rotation.

The frequency of sampling must reflect the way in which measured values themselves change with time. A rapidly changing measured value will have to be sampled much more frequently than a slowly changing value, in order to reconstruct its true nature from the samples. Figure 2.5 illustrates the dangers of sampling too infrequently.

The sampling of the input channels, based on the A/D conversion and sampling time alone, can be very fast. However, it is important to ask: *what sampling speed is required for the average building services plant to be adequately controlled?* Shannon's sampling theorem can be referred to in order to determine the proper sampling interval. This states that, provided

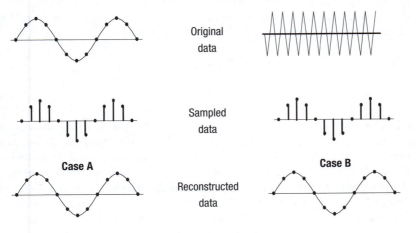

Figure 2.5 Examples of successful and unsuccessful sampling: A. Adequate (8 samples/cycle); B. Bad (<2 samples/cycle).

a signal contains no frequency component higher than f_{max}, the signal can be represented by (or reconstituted from) a set of sample values where the sampling frequency is at least twice f_{max}. In practice, sampling frequencies at least ten times the theoretical limit are often used.

In practice, to see how the valve or the rest of the plant is operating, selected sampled values from the input sensors can be stored or logged for later evaluation, most likely at the central station. A graph may be constructed by the central station software from this logged data. The outstation will have limited memory, and the sampling rate for this logged data will have to be carefully selected. When the memory becomes full, the later data will simply be allocated to memory space occupied by older data, and the older data will be overwritten and lost. However, the data will not be lost if it is regularly sent, or downloaded, to the central station for storage, where the storage is much larger. The frequency of downloading depends on the logging frequency.

2.3.2 Analogue to digital conversion

Conversion of analogue signals into discrete digitally coded values is fundamental to the application of digital computer techniques to industrial instrumentation systems. Unlike analogue values, digital values are discrete rather than continuous. Hence conversion requires that a search is made for the closest possible correspondence between the analogue value and its digital equivalent. A voltage proportional to the discrete value of each bit of the digital representation of the signal is generated and several voltages are added to obtain the analogue equivalent of the binary number. The process for an 8-bit A/D converter is detailed below.

The scaling factor, F, can be found by setting the maximum value of the analogue signal equal to the maximum of the digital representation. For instance, if the analogue range is 0 to 10 V, we have:

$$10 \text{ V} = \Sigma(128+64+32+16+8+4+2+1) \, F = 255 \, F$$

$$\text{so that, } F = 10/255 \, [=V_{range}/(2^8-1)] = 0.0392 \text{ V}$$

The least significant bit represents 0.0392 V while the most significant bit represents 5.0200 V. This scaling factor can be used to find the closest correspondence of digital to analogue measured value using an A/D converter of 8 bits. The maximum error in the digital equivalent is half the value of the least significant bit, taken as a fraction of the maximum of the measurement range. A simple equation, Equation (2–1), is often used to calculate the resolution ($R_{A/D}$) of the A/D converter, which is shown in the following example. The maximum error of the measurement caused by the A/D converter is half its resolution, assuming the sensor output is linear in the measurement range.

$$R_{A/D} = \frac{Measurement\ Band}{2^n - 1} \qquad\qquad (2\text{--}1)$$

The precision of the conversion increases and therefore the error band reduces with more bits used. For example, if a resolution better than 0.1 per cent is required, a minimum of 10 bits must be used. However, the greater the number of bits, the more expensive the A/D converter. Selection of the type of A/D converter therefore depends on the relative importance of precision and cost for the particular device concerned.

Example

A signal ranging from 0 to 10 V comes from a temperature sensor corresponding to the temperature ranging from –20°C to 80°C. What is the resolution of an 8-bit A/D converter?

Solution

Resolution may be defined as the change in the input signal for the digital output to change by the least significant bit (the last bit on the right of a binary number). An 8-bit word can be used to represent in binary the numbers from 0 to 255 (2^8 numbers including 0), so the resolution is:

$$10/(2^8-1) = 0.0392\ V = 39.2\ mV$$

or $\quad (80-(-20))/(2^8-1) = 0.392\ K$

(Note: the resolution of this A/D, 0.392 K, is too low for many applications, indicating the 8-bit length is not sufficient for many applications.)

2.4 Output unit

Having sensed the measurements via the sensors and the input section, the control outstation has to send a signal to control a valve or switch a device. This can be done simply by sending out a pulse. To move a valve by a certain amount a number of pulses will need to be sent to move the valve for a certain time. This is valve movement on an incremental basis. This approach used to be the most popular method for moving valves in the past.

However, the most common approach to conduct such modulating control actions, nowadays, is to use an analogue signal. If a valve needs to be moved to a definite position, a corresponding analogue signal is generated to move it to the defined position. In this case a digital to analogue (D/A) converter is required to generate the analogue signal. The resolution of the D/A converter can be determined by the formula, given in Equation (2–2), similar to that used for an A/D converter.

$$R_{D/A} = \frac{Output\ Band}{2^n - 1} \tag{2-2}$$

The output signals from the BAS control stations or D/A converters are control signals only. If large voltage or power, AC and DC, is needed to drive a device, a voltage or power amplifier is needed. Opto-isolation (transferring a signal via a short optical path, thus keeping the elements of the circuit electrically isolated) is often included in the output unit to protect the low-voltage microprocessor signal circuitry from any larger voltages. This prevents damage caused by the powerful driving circuits.

2.5 Processor operation and software

2.5.1 Machine and assembler languages

Different levels of language can be used to communicate instructions to a computing system. Typical examples are as follows:

- *machine language (expressed in machine code);*
- *assembler language (expressed in mnemonics);*
- *high-level languages (expressed in codes closely related to English words).*

An instruction consisting of sequential executions for the processor is represented by a unique set of binary characters (e.g. 010011000), which the machine can recognise as code for that instruction. A program might be prepared directly by the programmer writing instructions in this form, which is called *machine language*. But it is very difficult for people to use such language and therefore rare for programming to be carried out at this level today, although the processor can only recognize the instructions at this level.

Compared to machine language, people find *assembler language* more understandable, with each instruction being typed as a set of alphabetic characters known as a mnemonics. This enables the programmer to type the program and to understand the instructions typed. The mnemonics have to be translated into machine language which the processor can understand.

The following assembler program contains the instructions to take two units of data in the memory (address: 2240H, 2241H), add them and store the result to the memory (address: 2242H) for the processor Z-80 (a processor popularly used in the control industry during the 1980s):

LD A, (2240H)

LD B, A

LD A, (2241H)

ADD A, B

LD (2242H), A

HALT

However, these days most programs are written in ***high-level languages***, in which a typed word or set of words (closely related to words of the English language) describes an operation to be carried out on data. Each instruction at this level may require execution of several assembly-level instructions. High-level language programs must be compiled into assembler codes before being translated into machine codes to be run by the processor. Examples of high-level languages are Fortran, Basic and C (C is also known as a middle-level language).

Each single assembly-level instruction is implemented in a sequence of micro-operations or steps. In its turn, each micro-operation takes place typically in three stages, each initiated by a clock pulse:

1   The control unit sets up the control lines for the micro-operation.
2   The data or address is set onto the bus.
3   The operation is executed.

2.5.2 High-level languages

High-level languages consist of sets of instructions, each doing much more than a single instruction from the processor instruction set. Each instruction in a high-level language does something which is much more easily understood by the ordinary person but cannot be understood by the processor directly. Therefore, a program written in a high-level language has to be compiled. Compilation is the process of stringing together all the individual programs of assembly-level instructions, each representing a single 'high-level' instruction, to produce a program comprising a list of instructions which the processor is able to execute. The process of compilation is carried out by the processor, using a special program of assembly-level instructions, called a compiler.

Execution of the compiler program, which treats the high-level instructions as data, translates source code into assembly instructions or object code. The object code is in mnemonic form and must be translated into machine code before it can be executed. This translation process, like compilation, is carried out by another special machine-level program known as an assembler. Compilers and assemblers are special programs which generate assembly code (the compiler) and then machine code (the assembler) for execution by a particular processor.

To make it easy for a human operator to get the processor to carry out specific tasks, a special sort of high-level language is often produced, providing a set of 'instructions' each of which comprises a small program even at the normal high level. This set is automatically loaded into memory and each instruction can be invoked by an operator typing the appropriate mnemonic code together with any necessary data. A good example of this is the DOS or Windows 'environment' supplied with personal computers.

For a highly specialised area of operation, such as the programming of BAS control stations, these specialised but powerful instructions make it much easier for the user to configure his or her own system compared with the case if he or she had to write instructions in a generalised high-level language. Such a 'software environment' or 'programming environment' is created for the particular specialised systems.

A modern BAS is usually provided with some person–machine interface. One function of the interface is to provide the services of compiling and assembling. The interface allows the programmer to write the BAS application software in a general high-level language or even by simpler means specially designed. The BAS programming language and programming environment will be discussed further in Chapter 3.

2.6 Sensors

Sensors form a vital component of any control system. Sophistication in the computing and software functions cannot compensate for inaccurate information provided by poor-quality or inappropriately mounted sensors. The types of sensor available for use in building control systems are reviewed and guidance on selection and installation is given below.

2.6.1 Basic categories

The word sensor is generally used rather loosely to cover all processes between the measured variable and the input to the control module. According to function, a sensor may be broken down into three elements:

- *Sensing element:* a component that undergoes a measurable change, such as voltage or electrical resistance, in response to a change in the variable to be measured.
- *Transducer:* an active device that produces an electrical signal which is a function of the change in the sensing element.
- *Transmitter:* a device that produces an electrical signal which is a standardised function of the change in the physical variable and which can be used as an input to the control module.

In practice, the functions of the transducer and transmitter are often combined. Their combined function may be referred to as signal conditioning, which may also include filtering to remove noise, averaging over time, or linearization. In some systems, the sensing element may be connected directly to the controller, e.g. thermistor, where the signal conditioning takes place inside the controller module.

The various combinations are categorised as follows. *Status sensors* produce a binary (on/off) output depending on whether the measured variable is above or below a threshold. The sensors can be mechanical devices, where a physical movement of the sensing element causes switch contacts to open or close. Typical devices are thermostats, pressure switches and motion detectors. The output may be connected to a digital input of a controller for status reporting or software interlock purposes.

Analogue sensors convert the value of the measured variable into an electrical signal which is input to other devices for measurement and control purposes. Analogue sensors may be categorized into:

* *Passive sensors* which consist of the sensing element only and do not contain a transducer. All signal conditioning is carried out in the controller to which it is connected. Examples include resistance-type temperature sensors. No power supplies are required and the sensor is connected via field wiring directly to an analogue input on a controller.
* *Active sensors* which incorporate signal conditioning within the sensor device and include a transmitter which converts the measured value to an industry-standard electrical signal for connection via field wiring to an analogue input on the controller.

One such standard electrical signal form is a 4–20 mA signal which requires only a two-wire connection and is commonly used in process-control applications. Another typical form of signal used is a 0–10 V DC signal, which has widespread applications in HVAC systems. Other forms of signals are *voltage-free contact*, which is used for status indication, and *pulse*, which is used for energy and flow measurement.

Recent intelligent sensors contain a microprocessor which converts the measured value or status of the measured variable into a digitally encoded signal for direct communication over a network for onward transmission to other intelligent devices for control and measurement purposes. In addition, an intelligent sensor may carry out additional data processing before transmitting the value, such as checking upper and lower bounds, calibration and compensation functions, and calculating derived values, e.g. enthalpy.

2.6.2 General technical specifications

The correct selection and location of sensors is essential to achieving the required performance from any control system. Sensor problems are the most frequent cause of control system malfunctions. A poor-quality sensor may suffer from drift or early failure, resulting in poor control and high maintenance costs.

Accuracy: The claimed accuracy for a sensor does not guarantee that the same accuracy will be achieved at the controller or BMS supervisor, or that it will be maintained over the operating life of the sensor. The accuracy of the overall measurement system depends on many factors including: *accuracy of the sensing element, sensitivity of sensor element, insensitivity of sensor element to interacting variables, stability, hysteresis, mounting, signal conditioning, and A/D conversion.*

Speed of response: Sensors need to respond sufficiently fast to changes in the measured variable so that stable and accurate control can be maintained. The speed of response is characterised by the time constant T, which is the indication of the time taken for the signal output to follow the change of the measured variable from one level to another level. The time constant of a sensor in practice includes the effects of its housing, the manner of mounting and the nature of the medium being measured. There may be additional delays introduced by the measurement system. For instance, the scanning rate of the controller limits the speed of response of the system to a change in the measured variable. Increasing the relative speed of the fluid flowing past the sensor reduces the time constant.

The time constant of the sensor should be considered in relation to the rest of the controlled system. Too low a time constant may give problems if short-term fluctuations in the measured variable give rise to unwanted control action. This may be dealt with by the controller software, typically by incorporating an averaging function to extend the time constant. Too high a time constant may mean that the control system will respond too slowly to changes in the controlled variable. This is difficult for the controller software to compensate for.

2.7 Actuators

An actuator responds to the output signal from a controller and provides the mechanical action to operate the final control device, which is typically a valve, damper or switch. A wide range of actuators is available and the chosen actuator must address the following concerns:

* matching the mechanical requirements of the controlled device;
* matching the characteristics of the control system, especially the output signal of the controller;
* being suitable for its operating environment.

2.7.1 Electrical actuators

An electrical actuator requires at least the following connections:

* Power to power the driver, which may be 220 V AC, 24V AC, 24 V DC.

- Control signal from the controller, where 0–10 V and 4–20 mA DC are the most common.

The actuator may be fully modulating, where the position of the actuator is proportional to the control signal, or tristate, where the motor may be driven in either direction or stopped. Most actuators exhibit some degree of hysteresis, and the relation between control signal and actuator position depends on the direction of travel. Some actuators have the facility to provide a positional feedback signal, indicating the actual position of the actuator.

2.7.2 Pneumatic actuators

Pneumatic actuators comprise a piston or diaphragm to which air pressure is applied to provide a linear displacement. A mechanical linkage is required where it is desired to produce a rotary movement, e.g. for damper control. The construction of the actuator and its method of connection to the valve or damper determine the direction of operation. Most pneumatic actuators are of the single-action type where the force on the diaphragm is opposed by a spring and the net force applied to the valve or damper is the difference between them. When the air pressure is removed the spring will return the valve to the selected extreme position and this may be used for fail-safe requirements.

Pneumatic controllers provide reliable and fast operation and are still used in the HVAC industry. For new installations dominated by direct digital control (DDC), pneumatic systems are now installed only in special situations. Where an existing pneumatic control system is being upgraded to DDC control, it is possible to retain pneumatic operation of the actuators by using hybrid electro-pneumatic transducers which use pneumatic power to provide the operating force, but whose position is controlled by a standard electronic signal.

References

Borer, J. R. and Reynolds, A. J. (1994) *Building Management and Communication Systems*, Uxbridge: Brunel University.

Butcher, K. and Yarham, R. (2000) *CIBSE Guide H – building control systems*. Oxford: Butterworth-Heinemann.

Dorf, R. C. and Bishop, R. H. (1998) *Modern Control Systems*, 8th edn, Menlo Park, California: Addison-Wesley.

3 Building automation systems

3.1 What is BAS?

Building automation system (BAS) is an umbrella term (and is also known as building management system, BMS). It is used to refer to a wide range of computerized building control systems, from special-purpose controllers, to standalone remote stations, to larger systems including central computer stations and printers. As discussed earlier, BAS is one of the major intelligent building systems.

A BAS comprises several subsystems which are connected in various ways to form a complete system. The system has to be designed and engineered around the building itself to serve the services systems for which it is intended. Consequently, although the component parts used may be identical, no two systems are the same, unless they are applied to identical buildings with identical services and identical uses.

Building services include HVAC systems, electrical systems, lighting systems, fire systems and security systems and lift systems. In industrial buildings they may also include the compressed air, steam and hot water systems used for the manufacturing process. A BAS may be used to monitor, control and manage all or just some of these services. There are good reasons and ultimate objectives in investing considerable sums of money in this way. These will vary, depending on the use of the building and the way the building is managed as well as the relationship between the value of the end product and the cost of operating the building. It may also depend on the level of sophistication of the building services and their capital cost. The main typical benefits of having a BAS are discussed below.

3.1.1 Increased reliability of plant and services

The objectives of system operation and maintenance are to ensure the plant runs properly without breakdowns and to preserve efficient operation. Failure of a component almost always results in a more expensive repair or replacement than would have been necessary with timely periodic attention. Furthermore, the breakdown of certain equipment interrupts the service provided by the environmental system with resultant inconvenience to occupants and/or extra cost to the owner.

A BAS can make a significant contribution towards guaranteeing the operation by monitoring the system continuously and providing preventative maintenance. Typical examples are equipment alerts when the predetermined operating time has been reached and in the case of equipment performance having been degraded to a certain level.

3.1.2 Reduced operating costs

One of the major expenses in operating a building is the cost of energy required for heating, air-conditioning and illuminating the space. A key function of the BAS is to reduce the energy costs as much as practically possible. Typical examples of this are programmed start/stop, duty cycling, set-point reset and chiller optimizations.

The personnel used to maintain a building and its services is a significant portion of the overall operating costs nowadays due to increased remuneration costs and the increased sophistication of modern building services systems. The contribution which a BAS provides to reducing manpower requirements can have a major effect on the annual cost of running a building.

All types of buildings are candidates for some kind of energy-saving system. If the only reason for installing a system is to save energy, it is referred to as an energy management and control system (EMCS) or building energy management system (BEMS) rather than a BAS or BMS. Therefore, an EMCS or BEMS is normally considered as part of the BAS or BMS. EMCS or BEMS can be considered as the monitoring and control systems of building services systems that have significant contributions to the energy consumption of buildings.

3.1.3 Building management

BAS provides the most cost-effective means for staff to manage the building. This means monitoring the conditions and services and maintaining them at the required level at all times. It also means being able to respond quickly and efficiently to changes in function patterns and use of space. Detailed building-management functions are explained in Section 3.4.

3.1.4 Enhancing staff productivity

A BAS can also provide benefits which are less tangible and therefore difficult to measure. These include increased efficiency of personnel because of improved environmental conditions. Improved morale and job satisfaction of maintenance personnel, who are able to spend more time preventing things from going wrong and less time in 'fire fighting', can be another intangible benefit.

3.1.5 Protection of people and equipment

Inherent to the BAS is a communication network that extends throughout the building or complex of buildings. This same communication system can be put to work sending alarms to an operator or security service in the event of smoke, fire, intrusion or situations that could possibly damage equipment.

In addition, the BAS can also assist in other security measures. For example, it can control access to itself by providing the building manager with the capability of granting different levels of access to various staff members. The BAS can help guard against intrusion in the building by utilizing card access, by controlling and monitoring specified areas of the building, and by assuring that the rounds of security patrollers match a predetermined schedule.

3.2 The progress of BAS

Traditionally, building environmental control has been performed on mechanical equipment through pneumatic or electro-mechanical devices. With the widespread acceptance of DDC technology and use of microprocessor-based systems, building automation systems are replacing traditional controls and serving as the primary control systems. In the current building marketplace, more building services systems have 'built-in' control components. Typical examples are chillers with a control panel and VAV terminals integrated with control components.

Description of the overall development of BAS can cover its evolution since the early 1940s. The progress of BAS may be highlighted by the following stages or generations of building automation systems:

- *centralized control and monitoring panel;*
- *computerized centralized control and monitoring panel;*
- *BAS with data-gathering panel (DGP) based on minicomputer;*
- *microprocessor-based BAS using LAN;*
- *open BAS compatible with Internet/intranet.*

3.2.1 Pre-BAS stage: centralized control and monitoring panel

The 'pre-BAS stage' is labelled as such because the systems were not yet fully realized. Different technologies were introduced progressively at this time. Basically, the centralized control and monitoring panels allowed operators to read the sensor readings and start/stop or reset systems at one central location without running between the actual sensor and switch locations distributed over the building (Figure 3.1). The number of sensors and switches was very limited compared with the modern BAS, which may involve thousands or even hundreds of thousands of input/output points.

Pneumatic centralization achieved by using pneumatic sensor-transmitters permitting local indication and remote signal with a receiver-controller was

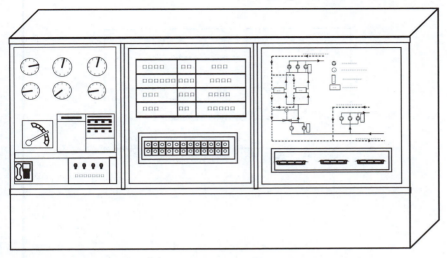

Figure 3.1 Centralized control and monitoring panel.

a technology introduced at this stage. The other technology introduced was signal amplification allowing an air signal to remain constant throughout its passage in one of the bundle of plastic tubes between the device and its controller. Consequently, the number of local control panels in a building could be reduced to a single centre.

The introduction of electronic sensors and analogue control loops resulted in a hard-wired centralized control centre. The problem faced at the early stage when using electrical technologies was excessive wiring for electrical signal transmission and therefore the high cost of installation. The introduction of electromechanical multiplexing systems in the 1960s reduced the number of wires, which resulted in reduction of installation costs and maintenance. The wires were reduced from hundreds to a few dozen wires per multiplexer. Commercial digital indication and logging systems were available on the central panels to permit the automatic recording of selected measurements.

3.2.2 First generation – computerized centralized control and monitoring panel

The first computerized building automation control centre was marketed late in the 1960s, following the development of the modern computer in the 1940s. A computer was connected to remote multiplexers and control panels, allowing all messages, sensors and devices to communicate through a coaxial cable or a two-wire digital transmission. The capability to address all points on the system provided operators with much useful information. The system (Figure 3.2) provided schedule programming of controllable

Central Computer

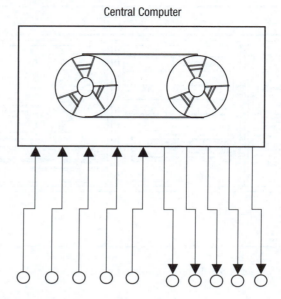

Figure 3.2 Computerized control and monitoring system.

devices, automatic reset of analogue outputs, high and low alarm limits, and reports.

The system of this generation was very expensive and not easy to use. Due to the high cost of hardware, disc storage was rare. Programs were loaded manually through a tape reader and it was very hard to change the programs. The BAS at this stage had low reliability as the entire system was based on a single central computer, and also involved excessive wiring. This generation of computerized BAS had few applications and it was quickly upgraded to a new generation of BAS.

3.2.3 Second generation – BAS based on minicomputer using data-gathering panel

The use of minicomputers, central processing units and programmable logic controllers (PLCs) in building automation systems increased dramatically in the 1970s. New application software packages were incorporated into their basic automation systems at an extra cost. Application packages for energy management were introduced, such as duty cycle, demand control, optimum start/stop, optimum temperature reset, day/night control and enthalpy control.

During the 1970s, the cost of hardware began to decrease dramatically. Computers became sufficiently practical to be used for common applications by non-specialist users. Compared with the computer systems of earlier stages, systems became user-friendly. It was much easier to program the

Minicomputer

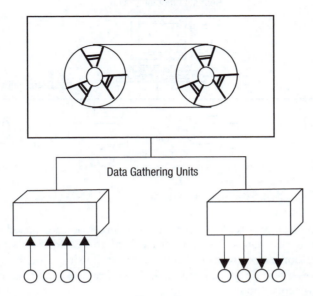

Figure 3.3 Building management system based on minicomputer using data-gathering panel.

systems and generate new databases. The use of keyboards and other new hardware provided users with a convenient primary human–machine interface (HMI) with the computers and building automation systems. The data and information collected by the computer-based automation system could be printed on paper or displayed on a screen. Communication between the operator and the computer-based systems, as commonly exists today, was initiated. However, the cost of special software increased because it could not be tailor-made in a simple way. In most cases, a specially trained programmer was the only person capable of doing this type of work.

Another important advance was the use of 'data gathering units', which meant that the data collected by many sensors and control signals and sent to many control actuation devices could be transmitted via a few wires. It significantly reduced the cabling work and allowed the building automation systems to have an extended number of monitoring and control points as required by the industrial activity in the building.

3.2.4 Third generation – microprocessor-based BAS using LAN

The use of microprocessors and the personal computer (PC) revolutionized control industries, resulting in the birth of a new generation of BAS. The lower cost of microprocessors and chips was the principal reason for the development of new technology in building automation and management.

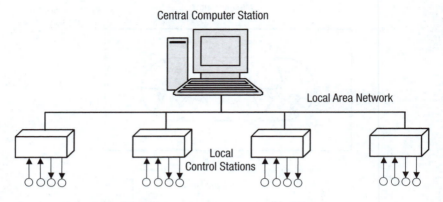

Figure 3.4 Typical microcomputer-based BAS using LAN.

The introduction and wide acceptance of microprocessor-based distributed direct digital control is the main feature of the BAS of this generation. Microprocessor-based control stations integrated using a local area network (LAN) represented the typical system architecture of BAS at this stage, which is still in existence today.

The use of a hard disc for data storage and loading of application programs provided great convenience in using and programming a BAS. A BAS normally had a central monitoring and management platform, running on a computer station, which was directly linked to remote control stations through a LAN. An important feature of BAS at (and after) this stage was the use of standalone but integrated microprocessor control stations to control individual plants. This allowed the BAS to have independent and distributed but integrated intelligence. It meant most of the control decisions could be handled locally, resulting in a significant increase in the reliability of BAS, while management and optimization could be done collectively.

The main problem for BAS at this stage was the incompatibility of different manufacturers' data communication protocols, message formats and information management as increased scale and functions required the use and integration of systems from different manufacturers. This problem was due to the fact that building automation systems did not comply with any standards commonly accepted.

3.2.5 Fourth generation – open BAS compatible with Internet/intranet

Since the 1980s, there has been a lot of effort put into developing and promoting standards in order to solve the incompatibility problems of building automation systems. The popular use of the Internet has also had a great impact on standardizing technologies used in the BAS industry. By the mid-1990s, open protocols and standard technologies began to be widely accepted and adopted in the industry. Many communication and software technologies commonly used by Internet/intranet or within the computing networking field have been adopted directly by the BAS industry.

The main feature of today's BAS can be summarized as follows. The use of open and standard communication protocols allows BA systems from different manufacturers to be integrated without much difficulty or effort. The use of IP and standard Internet/intranet technologies allows BAS to be integrated with enterprise computing networks conveniently. The convergence network provides a unified network platform for all information in buildings. BAS integration and information management can be achieved via the global Internet infrastructure.

3.2.6 Progress of BAS compared with computing technology

Figure 3.5 shows the progress of computing and BAS technologies and their interconnection. It is obvious that the evolution of BAS technology has followed the progress of computing technology due to the fact that BAS is actually the application of computing and IT technologies in building control and management. However, there was a clear boundary between the building automation systems and computing systems and networks in the first three stages, although the technologies originated from computing technologies. The typical BAS of the fourth generation is compatible with computing networks involving communication protocols and the means for information processing. There is no boundary between BAS and an intranet any more. The systems can be integrated easily at very large scales in terms of number of systems and geometry.

3.3 Programming and monitoring platforms and environment

3.3.1 Typical architecture of building automation systems and control stations

Figure 3.6 shows an example of a typical network architecture of a building automation system, although BAS are very different in terms of their scale and network configuration. In practical terms BAS, particularly large-scale BAS, may often involve more levels or layers of networks.

Field control networks typically connect the field control stations. The control stations are interfaced with the building services system via sensors, detectors and control actuation devices. Network control stations serve as the router/converter to integrate the field control networks into the management (higher level) network. They typically have relatively larger memory space and higher computation power. Network control stations may or may not have inputs and outputs for interfacing with building services systems directly. Field control networks typically have a lower communication speed.

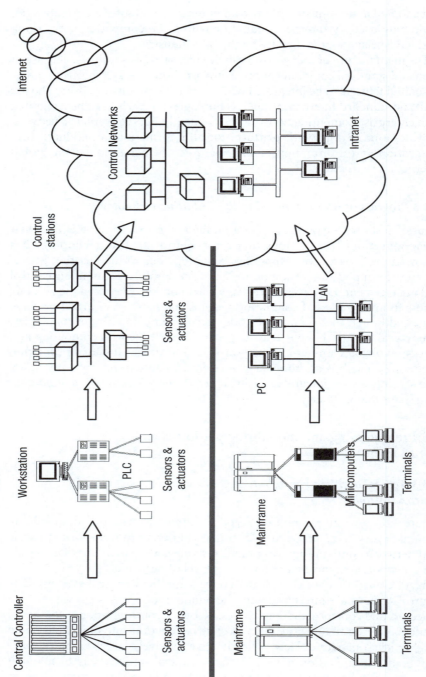

Figure 3.5 Progress of computing and BAS technologies and their interconnection.

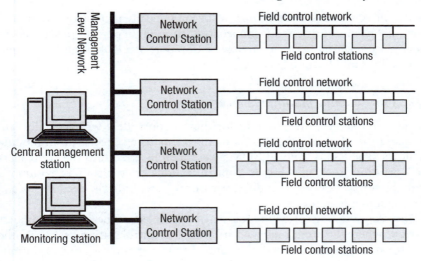

Figure 3.6 A typical network architecture of BAS.

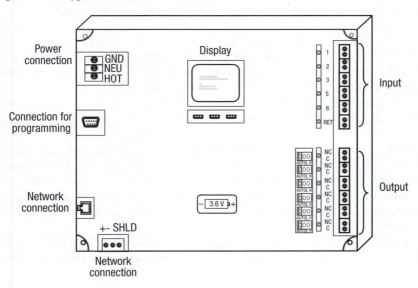

Figure 3.7 Example of typical configuration of field control stations.

The management level network involves computers and network control stations. The devices at this level communicate based on Ethernet, usually nowadays providing a very high communication speed. The computer stations connected at this level provide central management and information/ data storage as well as a platform and interface for operators.

Figure 3.7 shows a typical configuration and the function components of field control stations. A field control station typically involves a power connection, one or more network connections, connection port for programming (such as connection to the serial port or USB port of a PC), input and output connections and backup battery.

3.3.2 Programming platform and environment

In the early stages, the program for digital controllers was typically created and then fed into the ROMs or EPROMs using special programming facilities. The ROMs and EPROMs were then inserted into the controllers for operation to occur. Modern control stations are typically programmed with the support of user-friendly software tools, without the need for special programming facilities.

The programming software tools may be installed on the central computer stations where the programmers can configure control stations, program the control station and download the control software into the control stations from the central computer stations through the network. Some of the programming tools used on the central computer stations are separate software packages and others are combined with the monitoring or management software tools.

As the control stations governing different tasks integrated within the same BA network or BA system may be from different manufacturers, it is difficult to share the same programming tool on the central computer stations. Therefore, many control stations are provided with programming tools which typically run on PCs or notebook computers, linked to the control stations directly via the serial port, USB port, or the like, when programming them.

Programming a control station typically involves two main categories of tasks, configuring the control station, and developing and downloading the application program to the control station. Typical configuration tasks include: defining a station that runs as the server, defining channels and controllers, defining points and downloading the configuration database to the server.

The programming environments provided by the programming tools of different suppliers may be very different, and can be grouped into three categories:

1   graphic/symbolic format;
2   template or table format;
3   text format high-level language programming.

In all cases, the programming tools should provide the functions to enable conversion of the programs made by the programmers into machine language, which the microprocessor can execute, and download them into the control stations.

In a typical graphic-format programming environment, a graphic programming interface (tool) is provided with the function boxes representing commonly used function routines to conduct specific calculations, such as PID function, differential and so on. The control programs are made by simply selecting the proper function boxes and linking them correctly according to the control logic. The control program makes use of a measurement simply by making a link to the input channel and sends out a control decision by simply linking the particular outlet of the program flow chart to the dedicated output channel on the programming interface. Programming the control stations in this environment requires no special training in the programming language. The environment provides good flexibility for programming relatively complex control logic. However, it is not very efficient for making very complicated and sophisticated control programs.

When the control stations are specialized for certain control functions or control of certain building services systems where the control logic can be summarized into some generic format and the program for individual applications can be customized by defining or changing the parameters in the generic form or template, a more simpler programming form, template or table format programming tool can be used. Examples of control systems suitable to be programmed in this form are lighting control, security control and fire-detection systems. Limited programming freedom is provided when programming the control station in this environment.

Another means of programming is the text-format high-level-language programming. The languages might be similar to or even the same as one of the typical computer programming languages, such as Basic, C or Fortran. To suit the controller programming applications, some more rules might be introduced. Programming in this environment offers a great deal of freedom and flexibility to trained programmers. The advantages of programming a controller in this format become obvious when the control logic is very sophisticated. Programmers need much more training before they can use a particular programming tool in this format.

3.3.3 Management and monitoring platform and interface

Most modern BA systems provide a quite convenient and powerful user interface or human–machine interface for BAS network set-up, database set-up and controller set-up, as well as for system monitoring and management. Some systems share the same platform for both configuration and monitoring but others are designed on different platforms for monitoring and configuration. These days, many developers tend to separate two platforms due to the fact that the monitoring and management platform needs to manage a number of BAS subsystems from different manufacturers or using different standards. Building processes can be monitored by just using the system displays, which include alarm summary, point detail, trend and group displays. The typical display types are listed in Table 3.1.

Table 3.1 Typical display types

Display type	Description
Detail	Provides detailed information about a particular point. This information includes current values, scanning, history, etc.
Trend	Graphically displays changes in values, over time, of one or more variables. Trends can be displayed in several ways, including curves (graphs) and bar charts.
Group	Displays various types of information about related points on a single display.
Summary	Displays information, such as alarms and events, in list form. You can display more details about an item in the list by clicking on it.
Status	Displays detailed status information about system equipment, such as controllers and printers.

3.4 Building management functions

A building automation system is the high-technology tool or platform that expands and enhances the capabilities of those responsible for operations of a building. To better understand the potential impacts of a BAS, it may be helpful to look at the needs of the building operation and management which a BAS addresses. Typical functions provided by building automation systems include:

- *installation-management and control functions*;
- *energy-management functions (supervisory control)*;
- *risk-management functions*;
- *information-processing functions*;
- *facility-management functions*;
- *performance monitoring and diagnosis*;
- *maintenance management*.

3.4.1 Installation-management and control functions

One of the most frequently used terminologies in the field of computer control is direct digital control, because of the power this concept brings to the automation system. The increasing acceptance of DDC changed the overall nature of building control systems used in practice from traditional analogue systems to digital systems. Standalone control stations with DDC capacities have played significant roles in, and are increasingly important for, building automation.

Control functions of a BAS can be divided into two categories: local control (or installation-management and control) functions and supervisory control (or energy-management) functions. Local control functions are the basic control and automation that allow the building services systems to operate properly and provide adequate services. Local control functions can be further subdivided into two groups: sequencing control and process control. Sequencing control defines the order and conditions associated with bringing equipment online or moving it offline. The typical sequencing control in building systems includes chiller-sequencing control, pump-sequencing control, fan-sequencing control and lighting on/off control, among others. Process control is used to adjust the control variables to achieve well-defined process objectives in spite of disturbances, using measurements of state and/ or disturbance variables. Examples of typical process control in buildings are temperature control, air and water flow rate control and static pressure control. The most common feedback control function adopted for building processes is proportional-integral-derivative (PID) control. On/off control (or bang-bang control), step control and modulating control are the effective control-actuation schemes of local process control loops in building practice.

3.4.2 Energy-management functions (supervisory control)

For most BAS installations, savings brought about by improved energy management provide the economic justification for the purchase of the systems. The ways in which a BAS makes energy savings can be broadly grouped into two categories. The first is the savings which result through starting and stopping plants according to correct or optimal timing. The second is the savings which result through running plants in energy-efficient conditions, typically by setting the set-points of the local process controls at correct or optimal levels.

There is no better means of saving energy than that of turning off the energy-using equipment. Of course, you cannot turn off equipment that is needed constantly; we need to be able to turn off equipment without compromising the quality of services or the indoor environment. There are two approaches for starting and stopping equipment in an energy-efficient manner. They are called 'scheduled' and 'optimized' start/stop. In scheduled start/ stop, the HVAC equipment, lights and so on are turned on or off according to a combination of the clock and calendar. With an optimized start/stop program, the BAS assesses the existing conditions, anticipates conditions for the next several hours and decides when to start and stop the systems so that environmental conditions are provided during the complete building occupancy period with minimum energy use.

The control settings of the local controllers might be optimal and energy efficient when certain subsystems or certain subsystem performance criteria are considered. Supervisory control, often named optimal control, seeks to minimize or maximize a real function by systematically choosing the values of variables within allowed ranges. In the control of HVAC systems, for example, supervisory control aims at seeking the minimum energy input or operating cost to provide satisfactory indoor comfort and a healthy environment, taking into account the ever-changing indoor and outdoor conditions as well as the characteristics of HVAC systems. Compared to local control, supervisory control allows overall consideration of the system-level characteristics and interactions among all components and their associated variables.

3.4.3 Risk-management functions

In the same way that a BAS detects temperature and humidity conditions, it can also be used to detect fire or the presence of smoke. Fire safety integrated into BAS provides a greater degree of personnel safety than using two independent systems. The BAS is able, automatically, to close fire doors, close some air dampers and open others, start some fans and stop others and pressurize some parts of building with respect to others. This can help prevent the spread of fire and perhaps, more importantly, reduce the spread of smoke.

With the security system incorporated into the BAS, it almost always provides greater security and therefore reduces risk. Detection of someone trying to gain unauthorized entry is commonly by sensors on doors and windows. From the information reported to the central computer, the security officers can be made aware, not only that an intruder is present or is trying to gain entry, but also the intruder's location within the building. Access control differs from security monitoring since as the name suggests it is actually controlling who has access to a building or certain parts of a building.

3.4.4 Information-processing functions

Performing an economic evaluation of a large BAS is not a trivial assignment, nor can it be absolutely precise. The basic data needed for the economic evaluation is the cost of the BAS and the economic benefits that can be derived from the BAS. It is likely that the initial cost of the BAS can be estimated more accurately than the annual savings from energy conservation and other improvements. Although prediction of dollar savings attributable to energy conservation features and building management features is difficult, powerful systems provide energy monitoring and graph/table reporting, making estimation easier.

Engineers can directly access actual plant operating conditions through BAS to monitor energy use and energy cost, to carry out energy audits or to check performance using computer simulation techniques. With the support of BAS, a financial report can be produced with much less effort.

3.4.5 Fault detection and diagnosis, maintenance management, automatic commissioning

Fault detection and diagnosis (FDD) technologies, and smart maintenance schedule and automatic commissioning tools have followed in the wake of the development of information-processing functions.

Effective maintenance is a very important task of modern automation systems. It can extend equipment life, improve equipment availability and keep equipment in proper condition, maximize equipment efficiency, and consequently reduce the complete life cycle cost of the equipment. Effective maintenance is particularly important for intelligent buildings as modern IBs have many complex facilities, most of which have a direct relationship with the services quality and play a very significant role in the life cycle of the buildings.

Smart maintenance is proposed based on the monitoring data, which provides information on the equipment conditions. Conventional maintenance has been carried out according to schedule, which may be not suitable for all cases. Information-guided maintenance provides the service when it is needed, therefore saving manpower and reducing risk.

FDD technologies can be applied online or offline. An offline process is carried out based on the recorded monitoring data. Online technology is more advanced, and is able to detect and analyze faults while the system is running and produce a report concurrently. Automatic commissioning is a further development of FDD technology. Applying this technology it is possible not only to detect faults online but also to reflect the analysis result(s) to the system simultaneously for better control or even data recovery and fault-tolerant control.

3.4.6 Facility-management functions

As discussed earlier in Chapter 1 (Section 1.3), facilities management can have very broad definitions. FM professionals often consider almost all BAS functions as part of FM functions, and BASs are the systems used to achieve FM functions. Intelligent buildings need facilities management to define requirements, justify investment and deliver benefits. At the same time, facilities managers need intelligent buildings to control building performance, manage distributed services, adapt rapidly to changing requirements and provide crucial management information. From this point of view, the basis of facilities management is to ensure all the service equipment works properly. But from the viewpoint of building services engineers, FM functions mostly refer to the use of building spaces and facilities, including the economic effectiveness and financial aspects.

Intelligent buildings usually imply facilities management via building automation systems. Therefore, the facilities management of intelligent buildings requires the combination of an integrated BAS and the traditional information management system. Facility owners and managers require large amounts of data of various types for quality and efficient management. Typically, this information, such as management data of utilities, energy, maintenance, space, tenant and environmental compliance, is available and recorded on various computers or control stations.

Regarding the traditional information management system, up to now, in practice many facilities management systems (software tools) have been widely used in facilities management, such as NetFacilities, CAFMTools and ARCHIBUS/FM. The tools in this category are usually integrated information management platforms, often web-based, providing computer-based space management, move management, work-order administration, vendor-interaction management and other FM functions. They provide real-time collaboration platforms between facilities managers, maintenance staff, vendors, tenants and suppliers and others.

In practice, most of the facilities management systems are still single information management systems. They cannot retrieve data from integrated building automation systems, which are a huge data source. Future computer-aided facilities and maintenance management systems should provide more convenient and efficient management tools and exploit fully the advantage of integrated building automation systems in intelligent buildings.

References

Honeywell Inc. (1989) *Engineering Manual of Automatic Control for Commercial Buildings: heating, ventilating, air-conditioning*, Minneapolis, Minnesota: Honeywell Inc.

Scheepers, H. P. (1991) *Supporting Technology for Building Management Systems*, Woerden, The Netherlands: Uitgeverij De Spil BV.

4 Principles and technologies of local area networks

Integration is one of the main features of modern building automation systems and intelligent building systems. This integration is of the digital stations or devices (system integration) and the integration of control and management functions (function integration), while system integration provides a basis.

In a modern building, there might be a large number of digital stations or devices to be integrated. Local area networks are the primary choice for the integration of such large numbers of stations or devices in a building or within a short distance (say, a few kilometres). LANs have been used for data transmission among the stations or devices in the networks. However, it is also normal nowadays to transmit (digitalized) image signals and voice signals over a LAN.

This chapter introduces the basic principles of LAN, LAN topologies, the OSI reference model, LAN protocols, medium access methods and some typical LAN technologies for BAS/IB applications.

4.1 LAN characteristics

4.1.1 Wide area network (WAN) and local area network

WANs and LANs are distinguished on the basis of geographical distribution of the devices in the networks.

WANs provide communications over a long distance and have no geographical boundary. They rely upon the infrastructure, such as telephone networks and Internet backbone, which has already been installed and therefore provides worldwide coverage. Users do not therefore own their links, but instead pay a time-based charge to the services providers when using them.

WANs and LANs are used to link networks and digital devices separated by long or short distances together, but it is in fact hard to distinguish an exact geographic boundary between them. Typically, a LAN will be over a small geographic distance, normally a single building, or at most a campus. It will usually be privately owned (as opposed to being leased from telecommunication service providers) and these days it will normally have a data rate (measured in bits per second) of somewhere between a few Kbps and

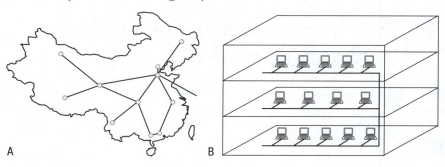

Figure 4.1 A. Wide area network (WAN); B. Local area network (LAN).

1,000 Mbps (1 Gbps). For computer LANs, it is normally between 10 Mbps and 1 Gbps, and the LANs of BAS and IB systems may range from a few Kbps to 1 Gbps.

The BA/IB and industrial automation networks may use various types of LANs. It is currently common to use an Ethernet (ranging from 10 Mbps to 1 Gbps) network at the higher level of the BA/IB networks. It will almost certainly be an Ethernet 802.3 network for computer networks today, although we should not forget that there are many network types. For example, many organizations still use Token Ring networks, particularly organizations demanding very high reliability.

4.1.2 Centralized networks and decentralized networks

Two general types of computer control networks can be distinguished: centralized networks and decentralized networks. These names relate to the operation of a network rather than the physical arrangement of the nodes and transmission links (where, a *node* refers to a computer-based station or device in a network). In a centralised network, there is a 'master' or 'boss' – a network controller (or central station) that controls the data transfers between all nodes in the network. Usually, all transfers must pass through the central node even if they are only exchanges of information between two peripheral nodes. In decentralized networks, all nodes have the same right to use network links and all are governed by the same rules, regardless of whether it is a very large or a very small station. There is no distinct network master as is the case in centralized networks, and control over the access to network links is distributed among all network nodes.

Comparing centralized and decentralized networks, one can find that centralized networks are less reliable due to the fact that they cannot operate if one of the network nodes, the central node, fails. But a failure of any one of a decentralized network's nodes does not affect other nodes. On the other hand, network protocol for a centralized network is simpler than a protocol for a decentralized network.

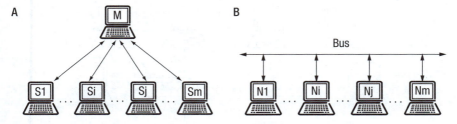

Figure 4.2 A. Centralized network; B. Decentralized network.

The architecture of a centralized network is illustrated in Figure 4.2A. Two types of network nodes are distinguished here: the master node (the network controller) and slave nodes. In this model, all messages are transferred through the master node. For instance, if the ith slave node wants to send a message to the jth slave node, the message will first travel from the ith slave to the master and then from the master to the jth slave.

An example of a decentralized network is presented in Figure 4.2B. In this network, a token-bus technique is used, in which a node has the right to transmit when it possesses a token. A token is a special, unique message that can be transmitted along the network link from one node to another. Once a node receives a token, the node assumes that it obtained the right to use the network link.

4.1.3 LAN topologies

The topology of a network is the manner in which each element of it connects to its partner both physically and logically. Over the years, several different methods of physical interconnection have evolved. Each of them has its own particular merits and shortcomings, and none can really be said to be absolutely superior to any other. Below, the three most popular topologies are introduced together with a brief description of their advantages and disadvantages.

Star topology: This is possibly the oldest of all topologies. As shown in Figure 4.3, there is a central hub through which all traffic flows. The obvious disadvantage of this system is that the hub itself is a single point of failure. If the hub fails, all stations connected to it will lose connectivity with the rest of the network. The advantages of this system are that it is simple to implement, and since all devices are connected to a single central point, the network can be monitored easily and efficiently. It does not mean that one cannot expand the network. When more devices need to be added, more hubs are simply connected as shown in Figure 4.3. This type of topology can be seen in just about all LAN types nowadays. However, Ethernet/IEEE 802.3 may use a physical star topology for interconnection, but logically the topology is a bus. Equally, Token Ring and FDDI (Fiber Distributed Data Interface) may

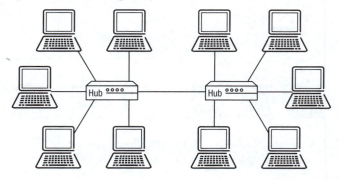

Figure 4.3 Star topology LAN.

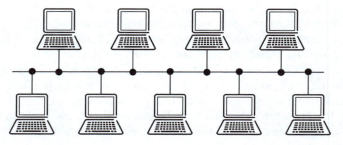

Figure 4.4 Bus topology LAN.

use a physical star topology to connect devices such as concentrators, but logically they operate as a ring.

Bus topology: The bus topology is most commonly found in Ethernet/ IEEE 802.3 (ISO/IEC 8802–3) environments, although it can also be found in other LANs such as ARCnet that uses the Token Passing Bus standard of ANSI 878.1, which is close to IEEE 802.4 (ISO/IEC 8802–4). Typically implemented in coaxial cable, and terminated at each end, this topology is simple and inexpensive to install, and equally simple and inexpensive to expand. These advantages may become its major disadvantage as well, since bus topology networks are often expanded without thought or planning, and they grow beyond their design limitations. In addition, any break in the cable renders the entire network inoperable. Figure 4.4 shows a simple example of a bus topology LAN in which either Ethernet/IEEE 802.3 or IEEE 802.4 could be implemented.

Ring topology: Figure 4.5 shows a LAN using a ring topology. This is normally associated with Token Ring and FDDI environments although, as said before, these LANs are often presented as star topology networks physically nowadays. Certainly, when Token Ring and FDDI environments are chosen, there is a relatively complex access protocol using an ordered method of node access. That means, with ring topologies, one has a robust

LAN. One of the reasons for this is that these environments can use dual, counter-rotating rings that can tolerate single cable breaks. The LAN access devices in a network using ring topology can simply detect the break and re-route the data over a backup path. The LAN then becomes self-healing. In addition, since node access is strictly controlled, the latency of the LAN can be predicted easily. Thus, stations take turns in accessing the LAN, and therefore networks employing this topology are generally deterministic, since performance can be predicted relatively accurately based upon the number of nodes involved.

The disadvantages associated with ring topologies are that the technologies they are employed for tend to be expensive to implement. This is due, in part, to the fact that the access protocols are more complex, and the devices are correspondingly more intelligent. Also, Token Ring and FDDI LANs have never achieved the same level of popularity that Ethernet/IEEE 802.3 enjoys. This means that access devices are not produced in the same quantities, and the prices remain considerably higher.

In terms of operation, a station that has data to transmit will wait for an invitation in the form of a special frame called a token. Once acquired, this special frame allows the station to transmit for a limited amount of time. As such, the station will now transmit its data to its downstream neighbour which will examine the frame, determine if it is addressed to it, and then pass it to its downstream neighbour. Where, on examination of the frame, a station finds that it is destined for itself, that station will copy the frame to its memory, mark on the frame that it has received the data, and pass the frame to its downstream neighbour as well. There are two reasons for this. First, the frame may be destined for more than one station. Second, by passing the frame from station to station, it will eventually arrive back at the station that sent it. The transmitting station can then check to see whether the data was received, and then remove the frame from the LAN.

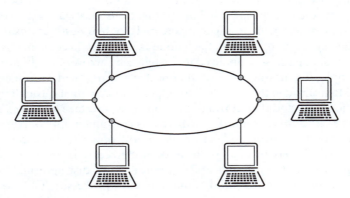

Figure 4.5 Ring topology LAN.

4.2 Network protocol and ISO Reference Model

To provide error-free and optimally convenient information transfers, the network operation is regulated by a set of rules and conventions called *network protocol*. The protocol defines connectors, cables, signals, data formats, and error-checking techniques as well as algorithms for network interfaces and nodes, allowing for standard (within a network) principles of message preparation, transfer and analysis on different levels of detail.

One of the overriding problems faced by the developers of the current generation of networks has been how to make the technology open and freely available. The mainframe environments in the past were based on proprietary solutions. The newly developing LANs, however, have had to be based on open standards to provide full vendor interoperability and therefore satisfy user demand. The days of end-users being locked into a single vendor have passed as the standards define everything from cables and connectors through to access methods and data format, allowing devices from difference vendors to be integrated easily.

4.2.1 Communication architecture

The term communication architecture refers to the fact that the overall networking task is partitioned in such a way that specific subtasks are performed by distinguishable network entities, architecture elements and information flow paths (communication links), and interfaces are established between them. The method by which messages are processed by architecture elements and transferred through information flow paths is called a network protocol. The protocol is closely related to the communication architecture. A protocol is implemented in the associated architecture. Communication architecture is also often referred to as protocol architecture.

Figure 4.6 represents a simple, generic model of network architecture including m network nodes, communication link and node/link interfaces. Messages are created and consumed at network nodes. The purpose of a network as a whole is to ensure reliable and fast transfer of messages from their sources (nodes at which messages are created) to their destinations (nodes at which messages are consumed). Each node can create and consume messages. Therefore, data paths – symbolized in the figure by arrows – are bidirectional. For example, if a message is created at node i and destined for node j, it has to travel first through the node/link interface at node i, then through the communication link, and then through the node/link interface at node j. Finally, it arrives at its destination node j, where it is consumed. Of course, another message may travel from node j to node i in the opposite direction.

In Figure 4.7, network communication architecture is presented in more detail. In this model, internal node operations are distinguished between application processor (AP) and communication processor (CP). Messages are created and consumed by APs, where CPs are responsible for message

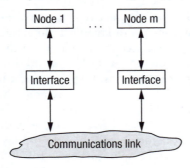

Figure 4.6 Communication architecture concept.

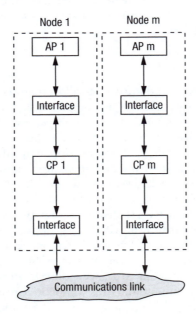

Figure 4.7 Three-layer architecture.

transfers from a source AP to a destination AP. The interface between the AP and CP can be regarded as a mailbox (namely, the application–communication [A/C] interface). The AP deposits messages to be transmitted and collects messages received. Messages deposited by the AP in a source node should be picked up by its CP and transferred, through the interface between the CP and communication link, and communication link to the CP in a destination node, and then be placed in its A/C interface and finally consumed by the AP. The same process applies in the opposite direction.

Such a partition of network operations into three parts, i.e. AP, CP and a communication link, brings very important benefits. Each of the parts can be defined and implemented separately, simplifying the development task.

Further benefits result from the fact that network models used in the BA profession follows the principle of layered architecture.

The benefit of layered architecture results from the fact that higher layers can use services provided by lower layers and do not have to deal with the details of lower layer operations. For example, a top-layer AP creates a message, places it in the A/C interface, and assumes that the message will be delivered error-free to the AP in a destination node. The AP can also assume that received messages placed in the A/C interface by the middle-layer CP are error-free. Eventually, from the viewpoint of the top-layer application processor, the network appears to be operating error-free. Of course, transmission errors do happen on the network, but they are handled by the middle or lowest layer, and top-layer APs are not involved. A commonly used approach to partition the tasks and functions into layers is to use the ISO Reference Model as described below.

4.2.2 OSI Reference Model

Development of hardware and software to allow many computers and stations to communicate on a network is a very complex task. It requires a lot of effort even if all elements of the network are developed by a single manufacturer. Manufacturers optimize their designs to make them cost-effective and thus more competitive. As a result, products developed by different manufacturers tend to be incompatible. If we attempt to create a network using computer and control stations from several vendors, the problem of integration becomes serious. The need to allow processor-based products offered by different manufacturers to communicate with each other on a network led the International Organization for Standardization (ISO) to propose a network architecture model referred to as the Open Systems Interconnection (OSI) Model. The goal of the ISO subcommittee developing the OSI Model was to provide a framework for networking standards, acceptable to all manufacturers, which would help them design unique and thus competitive products that could still communicate with compatible products offered by other vendors. The OSI Model is generic and applies to both WANs and LANs.

As explained in the previous section, the problem of complexity in computer network communication is best handled by using a layered-architecture approach, in which all networking functions are partitioned into several groups (called layers), in such a way that upper layers use services provided (functions performed) by lower layers. The OSI Model implements the layered architecture concept and defines a number of layers, the functions performed by each layer, and interlayer interfaces. Partition of all networking functions into layers is guided by two contradictory constraints. When more layers are used each becomes smaller and simpler. On the other hand, the use of many layers creates many interlayer interfaces, and the processing overhead necessary to handle additional interfaces may offset the benefits gained by layer simplification.

Developers and vendors can each produce component parts of the whole, yet provided that the interface between layers remains constant, there will be assured interoperability. Figure 4.8 shows the OSI Reference Model, and the layers on the basis of which many of the standards are developed.

The OSI Model partitions networking functions into seven layers, as shown in Figure 4.8. Once the OSI Reference Model was adopted, standards defining each of the OSI layers could be developed. For example, if a standard is established for the Data Link (DL) Layer, specifying its functions and upward and downward interfaces, any product that implements the standard in any way is compatible with other products that obey the same standards. Usually, a LAN protocol refers to the specifications of the first two layers, i.e. Physical Layer and Data Link Layer, which are technology-dependent layers. The upper layers, i.e. Application, Presentation, Session, Transport and Network Layers, are technology-independent layers.

Physical Layer: The Physical Layer specifies the electrical (and/or optical, or even radio) signalling, and the mechanical or physical connections applicable

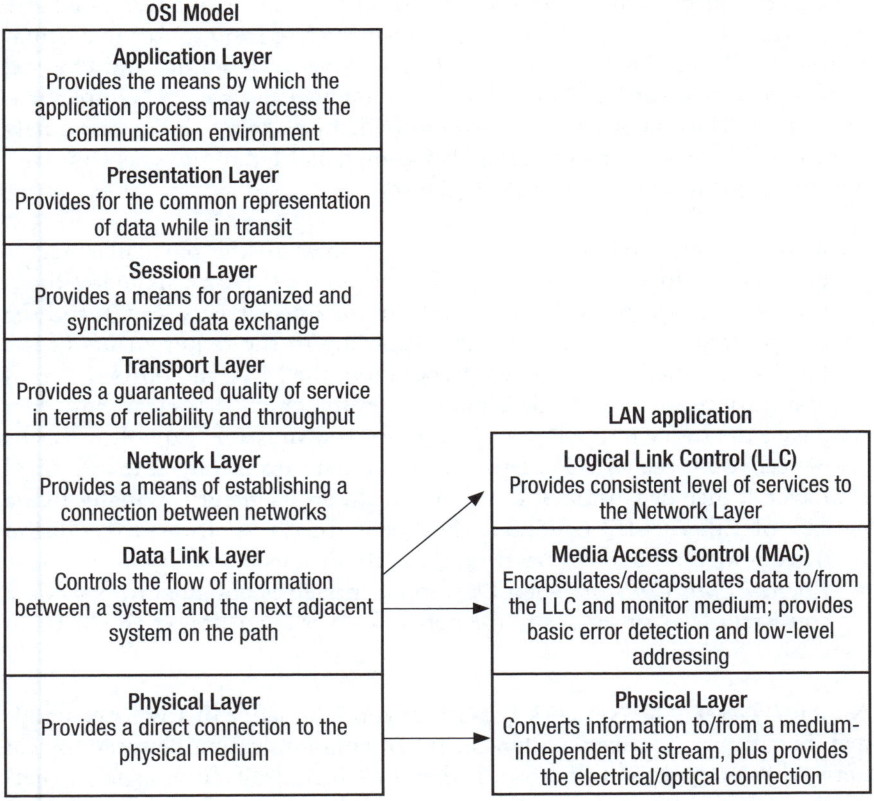

OSI Model

| **Application Layer** |
| Provides the means by which the application process may access the communication environment |

| **Presentation Layer** |
| Provides for the common representation of data while in transit |

| **Session Layer** |
| Provides a means for organized and synchronized data exchange |

| **Transport Layer** |
| Provides a guaranteed quality of service in terms of reliability and throughput |

LAN application

| **Network Layer** Provides a means of establishing a connection between networks | **Logical Link Control (LLC)** Provides consistent level of services to the Network Layer |

| **Data Link Layer** Controls the flow of information between a system and the next adjacent system on the path | **Media Access Control (MAC)** Encapsulates/decapsulates data to/from the LLC and monitor medium; provides basic error detection and low-level addressing |

| **Physical Layer** Provides a direct connection to the physical medium | **Physical Layer** Converts information to/from a medium-independent bit stream, plus provides the electrical/optical connection |

Figure 4.8 The ISO Reference Model.

to the medium type in use. It provides a physical path for electrical signals representing bits of transmitted information. It also defines the characteristics of these signals, such as voltage and current levels, frequencies and timing. It specifies the mechanical properties of network cables and connectors. From the point of view of upper layers, the Physical Layer ensures that streams of bits produced at the interface between the Data Link and Physical Layers by the transmitting node arrive at the interface between the Physical and Data Link Layers in all receiving nodes in a bus-topology LAN. The Physical Layer is the only real interconnection between network nodes.

Data Link Layer: The Data Link Layer defines rules for sharing the use of the Physical Layer among network nodes. On a bus-topology LAN, the medium (Physical Layer) is shared in time-division fashion. Information is transferred in addressed frames, one frame at a time. The format of these frames and the method by which a node decides when to transmit or accept a frame are defined. Two general types of frames are used. Data frames, which convey the upper-layer messages, are also referred to as packets. Other frames used by the Data Link Layer, such as token or acknowledge frames, are called control frames. Error-detection and error-correction techniques are used to ensure that packets are transferred error-free from source to destination nodes. From the point of view of upper layers, Data Link and Physical Layers provide error-free transmission of data packets. For LAN implementations, the Data Link Layer is further divided into two sub-layers, namely the Media Access Control (MAC; also known as Medium Access Control), and the Logical Link Control (LLC) layers.

- *Media Access Control:* The MAC is peculiar to the particular access method employed on the LAN. The MAC itself performs many functions such as the reception of data from the upper layers (LLC), and the encapsulation of this into frames according to the requirements of the LAN access method employed. In addition, the MAC monitors the communications channel to determine when the channel is clear, and then passes the frame to the Physical Layer for transmission. Equally, in terms of reception, the MAC accepts incoming data frames from the Physical Layer. It then decapsulates the data and checks its integrity. Finally, in the case of Ethernet/IEEE 802.3 LANs, the MAC is responsible for collision detection, and the recovery from these conditions.
- *Logical Link Control:* The LLC sub-layer itself is designed to provide a consistent level of service to the network layer regardless of the MAC in use.

Network Layer: The Network Layer is beyond the scope of any LAN standard, but it provides users with a means of communicating between logical networks (as opposed to LANs). It therefore includes facilities such as network routing, addressing and, in some cases, flow control.

Transport Layer: This layer provides the basic interface between the

Session Layer and the underlying network-dependent protocols. It is the layer that will typically provide for connection-oriented sessions which demand the exchange of data in an orderly and reliable manner. This reliability, which is not generally available at the Network Layer, is normally implemented as a sequence number/acknowledgment system. This then ensures that all data is received, and in an ordered manner.

Session Layer: The Session Layer mainly provides a method by which two systems may organize and synchronize their dialogue, and therefore manage the exchange of data between themselves. The Session Layer itself is possibly the most complex of the layers.

Presentation Layer: The Presentation Layer is concerned solely with the presentation of data while in transit. From the communications point of view, this layer is the simplest. Its function is to convert user messages from the form used by the Application Layer to that used by all lower layers. The purpose of message conversion (encoding) is to achieve data compression or security. At the upper interface of the Presentation Layer, data fields of messages have meaningful explicit form. Below the Presentation Layer, data fields of messages and packets are treated as meaningless envelopes, and their meaning does not influence their processing.

Application Layer: Contrary to its name, this layer does not represent the actual application, but instead is the application protocol. As such, the Application Layer provides the application itself with a gateway to the communications environment. Certainly, many application protocols do have applications of the same names, i.e., Telnet, FTP, etc., but this is not always the case. For example, mail applications would use the Simple Mail Transfer Protocol (SMTP), and the Post Office Protocol (POP), but application programs implementing them tend to have names without any relation with them, such as Microsoft Outlook. Similarly, web browsers use the Hypertext Transfer Protocol (HTTP), and the application programs using it are far more likely to address the tasks that they perform, such as Microsoft's Internet Explorer.

4.3 Medium access methods

4.3.1 An overview of LAN physical interface

All microprocessor operations are performed using strictly digital signals. Transistor–transistor logic (TTL) level signals used within microprocessors or control stations are typically at a voltage level between 2.4 V and 5 V representing the logical value 1, and a voltage level between 0 V and 0.8 V representing the logical value 0. Signals transmitted among the elements within the microcomputers or control stations are in the form of parallel digital signals, while signals transmitted among the network media are in the form of serial (analogue or digital) signals.

On the other hand, to transmit or receive data to and from a medium,

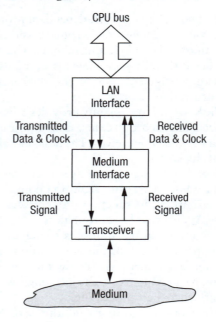

Figure 4.9 Physical interface.

some form of signal encoding must be used. This function is performed by the *medium interface*. Specific operations performed by the medium interface unit depend on the signalling and synchronization methods used. In general, it can be said that the medium interface provides bidirectional signal conditioning. In the transmit direction, streams of transmitted data bits are converted by a medium interface to electric signals appropriate for medium use. In the receive direction, a signal arriving from the medium is decoded, and the resulting stream of received data bits is produced. As shown in Figure 4.9, there is one more element between the medium interface and the medium itself, the *transceiver*. The role of the transceiver is to connect two unidirectional signal lines into one bidirectional line, which then can be tapped directly to the LAN cable bus.

As shown in Figure 4.9, the *LAN interface* operates between the CPU bus and medium interface unit. All LAN interface inputs and outputs have the form of pure digital signals with standard TTL levels. LAN interface typically implements a LAN coprocessor, which can operate on the same bus as the host CPU (main processor) and can execute instructions from main memory shared with the host CPU. The LAN interface functions include the following:

- All Data Link Layer operations, e.g. node-level addressing, error detection and medium-access control. In the token-bus protocol, medium-access control consists of token passing, network reconfiguration and token

recovery. In the CSMA/CD protocol, the LAN interface performs all collision-related operations, such as computation of the back-off time and transmission retry.
- Input and output of frames represented by bit streams to and from the medium, through the medium interface, at a fixed data rate. This function is particularly demanding in the receive direction. Bits arrive from the medium, through the medium interface, at a high, fixed data rate and cannot be slowed down. Frames represented by these bit streams have to be analyzed and accepted or rejected very rapidly.

Data frames (in both directions) as well as control and status information have to be exchanged between the LAN interface and CPU programs. The LAN interface implements the Data Link Layer, and higher protocol layers are implemented in the form of CPU programs. CPU programs operate on data located in main memory. Therefore, frames, control and status information have to be between the LAN interface and CPU programs through a shared area in the main memory.

4.3.2 Communication media

Communication medium is used to describe the physical path between transmitting and receiving devices in a communications network. Several types of media are used in electronic communication: cables (including power lines), radio waves, infrared and ultrasonic waves, and light waves. The most suitable media for LANs include twisted-pair and coaxial cables as well as optical-fibre links facilitating light-wave communication. Infrared and ultrasonic links are used for point-to-point links and, in general, are not implemented as LAN media.

The way in which network nodes (stations) are connected using communication media is referred to as network topology, which is described in Section 4.1.3. Another important concept in addition to network media and topology is *broadcast transmission*. This term applies to a bus topology with a twisted-pair or cable medium. In broadcast transmission, a signal injected into the medium from a transmitting node propagates through the entire medium and reaches all other nodes in the network nearly simultaneously. The benefit of the broadcast transmission is that network operation becomes simple, and no buffering or retransmission is necessary. However, this benefit is achieved at a cost. Since transmitted signals penetrate the entire medium, only one node can transmit to the medium at a time. If two or more nodes transmit simultaneously, their signals interfere with each other in the medium, and the information conveyed by the signals is lost. Such a situation is called a *collision*, and the network protocols for bus-topology LANs have to solve this problem of collision.

In both the twisted-pair and coaxial cables, shown in Figure 4.10, two conductors are used. One conductor is for the useful signal and the other

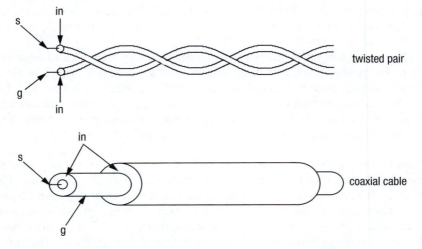

s: signal conductor, g: signal-ground conductor, in: insulation

Figure 4.10 Examples of network cables.

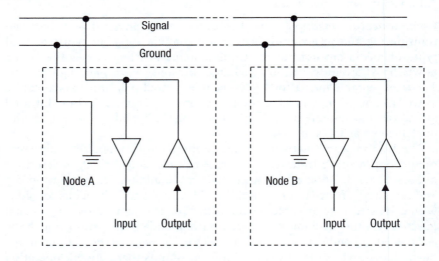

Figure 4.11 Node-to-node interconnection.

is used as reference or a signal ground. In coaxial cable, the role of signal-ground conductor is always played by the external cylindrical conductor. In both cables, conductors are insulated to provide electrical separation between each other and from any external circuits.

In bus topology, all network nodes are connected to the network medium (cable) in the same way. Figure 4.11 represents a model of such a connection, which is part of the function of the transceiver. The model is generalized and

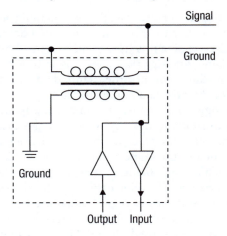

Figure 4.12 Galvanic separation of network nodes.

applies to both twisted-pair and coaxial cables. Each node is equipped with signal amplifiers, illustrated in the figure by triangles. Neither the twisted-pair nor the coaxial cable is a lossless medium. The signal is attenuated while it propagates through the medium. A node that transmits has to provide a signal of sufficient energy and thus needs an output amplifier. The receiving node has to be able to recognise the attenuated signal arriving through the medium and therefore it needs an input amplifier also. Since all nodes in a LAN can be transmitting or receiving, all are equipped with both output and input amplifiers.

Because LAN nodes can be located as far as a few kilometres away, galvanic separation provides better links between the nodes. This is achieved by using radio-frequency transformers, which separate input/output amplifiers from the medium in terms of ground potential but still allow high-frequency signals to pass through between node I/O and medium cables. Such an interconnection is illustrated in Figure 4.12.

4.3.3 Signal encoding and medium interface

Data to be transferred between LAN nodes exists in parallel digital form originally. Each message is a string of bits encoding some information. The interconnections between nodes in LANs allow only one bit of information to be transferred at a time, which is called *serial data communication*. Thus, a transfer of a message from one node to another must be achieved by sequentially transmitting all bits of the message from the source node to the destination node.

The source and destination nodes of a LAN are connected by a physical medium, such as a twisted-pair or coaxial cable. Physically, data bits are transferred in the form of electrical signals. Two major types of signals are used: analogue and digital signals. In analogue signalling, an analogue carrier signal, a sinusoidal wave, is used to convey encoded data. In digital signalling, a two-level, discrete signal is used. The benefit of analogue signalling lies in the fact that analogue signals are less susceptible to distortion due to attenuation in the medium. On the other hand, data encoding and decoding are much simpler for digital-signalling methods.

In the analogue signal case, only the amplitude is attenuated, but the waveform is not distorted and the original signal can easily be restored by amplitude amplification in the receiving node. In the digital signal case, not only the amplitude but also the shape of the waveform is distorted, and signal reconstruction is much more difficult. As a result, if a medium with high level of attenuation (e.g. long twisted-pair cable) has to be used, then analogue signalling would be the best choice. Digital signalling would be more suitable in LANs using media with a low level of attenuation.

When analogue signalling is used, digital data must somehow be encoded into an analogue signal before being sent. Two examples of such encoding methods are illustrated in Figure 4.13. In the *amplitude-shift keying* (ASK) method, an analogue carrier signal of a fixed frequency is used. If a bit of value '1' is to be transferred, the carrier waveform is transmitted. A lack of carrier signal means that a bit of value '0' is transmitted. Another method, requiring the use of a carrier signal that can assume two different frequencies,

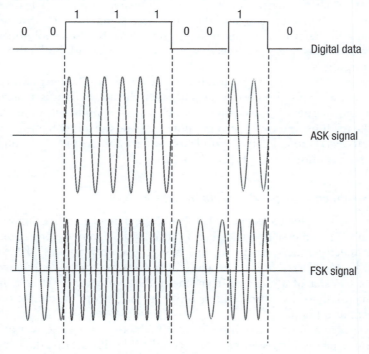

Figure 4.13 ASK and FSK analogue signal encoding for digital data.

is called the *frequency-shift-keying* (FSK) method. In this case, a logical value of '1' is represented by a carrier signal of frequency f_1 and logical value of '0' is represented by a carrier signal of frequency f_0.

The encoding of digital data into a digital signal is straightforward. Usually, the logical value '1' is represented by a positive voltage and logical value '0' is represented by a lower voltage. As a result, the signal level is always at one of the two voltage levels. However, no matter whether analogue or digital signalling is used, if two consecutive bits are identical in value (e.g. two '0s' or two '1s'), it is difficult to tell when the first one ends and the second begins, or how many '0s' or '1s' there are when a piece of signal is received.

To solve this problem, the signal transmitter (source node) and receiver (destination node) have to be synchronized; that is, they have to have the same perception of time. For example, a transmitter may provide a clock signal for the receiver in addition to a data signal. The use of an additional clock signal line is very efficient in solving the synchronization problem, but it requires an additional line connecting the transmitter and receiver. To avoid the use of a clock line while keeping the transmitter and receiver synchronized, two methods of data transmission might be used, *asynchronous* and *self-clocked*. Asynchronous transmission is used for lower data rates, whereas high-data-rate transmission is more efficient using the self-clocked method.

Asynchronous transmission uses start and stop bits to specify the beginning and end of each character of transmitted data. These additional bits provide the timing or synchronization for the connection by indicating when a complete character has been sent or received. Thus, timing for each character begins with the start bit and ends with the stop bit.

Manchester encoding of digital signals is a typical method used in self-clocked transmission, a type of method in the synchronous transmission category, as illustrated Figure 4.14. In this case, the receiver's clock is resynchronized on every bit transmission and bit streams of any length can be

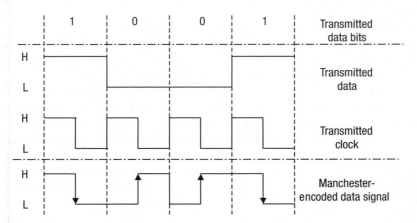

Figure 4.14 Manchester encoding (self-clocked transmission).

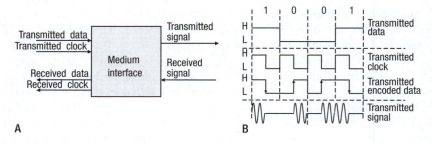

Figure 4.15 Medium interface using Manchester encoding method and ASK analogue
signals: A. Input and output lines; B. Encoding of transmitted data.

sent. The self-clock of the data signal is achieved by assuring an *H-to-L* or
L-to-H transition in the middle of every bit time. These transitions are used
to resynchronize the receiver's clock. When encoding the data bits, a '0' bit
is represented by *L* followed by *H*. A '1' bit is represented by *H* followed by
L. If no information is sent, there are no transitions on the data line at all.
The coding and decoding processes are conducted by the medium interface
in both directions. Generally, BAS engineers and users will not need to have
details of the coding and decoding rules at this level.

An example of a medium interface using ASK analogue signalling and
Manchester encoding is shown in Figure 4.15. On the digital hardware (LAN
interface) side, pure digital signalling with TTL signal levels and an additional
clock-signal line for each data line is used.

4.3.4 Examples of medium access methods

On the Data Link Layer in the communication model, there are several ways
of organizing this medium access. CSMA/CD and token passing are the typical methods used.

CSMA/CD: CSMA/CD is a distributed principle which is used in Ethernet;
that is, in a bus/tree topology (Figure 4.16). CSMA/CD stands for *carrier
sense multiple access/collision detect*. 'Carrier sense' means that a device that
wishes to send on the medium senses whether there are other devices sending
a carrier at the same time. The CSMA/CD is a contentious method, in which
network nodes compete for the right to use the medium. If the medium is in
use, the station does not send, but waits until the medium is free.

When the medium is free, the device waits for a random time and if the
medium is still free, the device starts sending its frame onto the medium.
Usually, it works, but there is some chance of collisions. In the event of a
collision, the nodes involved stop sending and try again after a random wait.
In this way, many nodes can use the same medium.

It may seem strange that the nodes wait a while first before sending when
the medium becomes free. The reason is that when the medium is busy there

may be more than one device waiting to use the medium. If every device starts sending as soon as the medium becomes free, a collision is highly probable. If a random wait is included, the devices will not all start sending at the same time and collisions can be avoided by the devices checking the medium once more before sending.

Token passing: The principle of token passing can be used on both ring and bus networks. This simple principle is based on defining a *token* in the form of a bit pattern, with 8 bits for example. Tokens circulate between the nodes if the network is a ring. If the network is a bus, the tokens 'circulate' in a particular sequence. Using a baton in a relay race as an analogy, only the one who 'is holding' the baton at a given instant can use the medium. This ensures that only one recipient uses the medium at a time. The principle has advantages over the CSMA/CD collision protocol when the network load is high. Administration of token passing is the same whatever the load, and the medium can be exploited to the full. On the other hand, the efficiency of CSMA/CD declines sharply when there are many collisions and the frames have to be resent.

Generally, token passing offers the ability to prioritize traffic. The principle

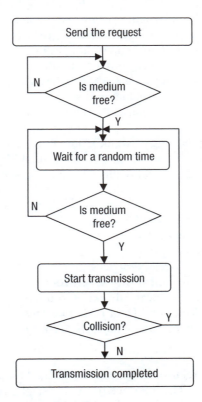

Figure 4.16 Logic scheme of CSMA/CD in accessing medium.

can also guarantee a maximum waiting time before a station can send. This has sometimes proved crucial when there have been requirements to process real-time data traffic.

4.4 An overview of LAN standards

In this section several of the standards are described that usually refer to local area networks, although according to the model described, it is actually the Data Link Layer (with the term 'local area network' being used for the Data Link Layer).

The standards for these LANs were developed mainly by the Institute of Electrical and Electronic Engineers (IEEE). IEEE originally developed standards for the speed range 1–20 Mbps, with the designation IEEE 802.x. IEEE 802 networks are '*peer-to-peer*' networks on a shared medium. Peer-to-peer means that the network connects equal parties (nodes). The fact that the parties are equal on the Data Link Layer means that the principle governing access to the medium is distributed equally between the parties. It is not the case that one device allocates network access to the others.

This principle of equality makes it possible to establish alternative point-to-point connections without switching elements. If you have computers A, B, C and D on a network, you may want to be able to have connections between A and B, between A and C, between D and B, or other arbitrary connections. Every computer can talk to every other computer.

The concerns for building up a LAN are, basically: topologies, medium access methods, transmission media and speeds. A series of standards was therefore created for different needs on these, but with a common interface at the top.

The top layer is called Logical Link Control. It was standardized by the subcommittee: 802.2. The layer below LLC is called Media Access Control, and was standardized by the following subcommittees as the protocol standards at the MAC layer:

802.3   CSMA/CD Networks
802.4   Token Bus Networks
802.5   Token Ring Networks
802.6   Metropolitan Area Networks
802.7   Broadband Technical Advisory Group
802.8   Fibre Optic Technical Advisory Group
802.9   Integrated Voice and Data LAN Interface
802.10   Standard for Interoperable LAN Security
802.11   Wireless LAN

Table 4.1 shows the relationship between LLC, the common MAC layer and the Physical Layer. The interface to the Physical Layer conforms to the definition of the Physical Layer in the OSI Model.

Table 4.1 LAN standards of IEEE 802 series

Data Link Layer		Physical Layer
Logical Link Control	*Media Access Control*	
802.2 LLC Logical Link Control	CSMA/CD 802.3	Baseband coaxial 10/100 Mbps
		Twisted pair 1, 10, 100 Mbps
		Broadband coaxial 10 Mbps
		Optical fibre 100/1000 Mbps
	Token Bus 802.4	Broadband coaxial 1, 5, 10 Mbps
		Carrier band 1, 5, 10 Mbps
		Optical fibre 5, 10, 20 Mbps
	Token Ring 802.5	Twisted pair 4, 16 Mbps
		Unshielded twisted pair 4 Mbps
	FDDI (Token Ring)	Optical fibre 100 Mbps

IEEE 802.2 LLC: The main task of LLC is to even out differences in the various types of local area network, so that there is no need for higher-level software to take these differences into account. However, the LLC layer is seldom used. In most cases this sub-layer is omitted in order to increase efficiency. In such cases, the layer above, e.g. IP, works directly with the MAC layer. So the software must be configured specially for the underlying layer used in each case. In most cases this is the IEEE 802.3 Ethernet.

4.5 Examples of LAN technologies in applications

4.5.1 Ethernet (IEEE 802.3)

The term *Ethernet* refers to the family of LAN products covered by the IEEE 802.3 standard that defines what is commonly known as the CSMA/CD protocol. Three data rates are currently defined for operation over optical-fibre and twisted-pair cables: 10 Mbps (10Base-T Ethernet), 100 Mbps (Fast Ethernet) and 1 Gbps (Gigabit Ethernet). Ethernet of even higher speed, such as 10 Gigabit Ethernet, is under development.

Other technologies and protocols have been promoted as likely replacements, but the market has made its selection. Ethernet has survived as the major LAN technology in PC and workstation networks. It is also one of the most popularly used LAN technologies in BA networks nowadays. Ethernet protocol has the following characteristics. It:

Table 4.2 Ethernet technical specifications

Specification	Characteristics
Network standard	ISO 8802–3/IEEE 802.3 serial
Protocol	Standard for layer 1 and 2
Architecture	Bus or star topology
Transmission speed	10 Mbps – 100 Mbps, 1 Gbps
Network node capacity	48 bits addressing (normally, 254 nodes per subnet)
Network length	100 m (10Base-T) to 10 km (fibre)
Physical path cabling	Twisted pair, coaxial, optical fibre
Media Access Control (MAC) method	CSMA/CD

- is easy to understand, implement, manage, and maintain;
- allows low-cost network implementations;
- provides extensive topological flexibility for network installation;
- guarantees successful interconnection and operation of standards-compliant products, regardless of manufacturer.

Ethernet uses a collision-detection mode of LAN communications. Since there is no token associated with this LAN, nodes are free to send messages whenever there is no other activity on the link. Messages sent by two nodes at the same time will collide. The message will then be lost and the nodes will wait for a specified period of time before attempting to resend. The technical specifications of Ethernet are shown in Table 4.2.

4.5.2 ARCnet (ANSI Standard 878.1)

This is a token-passing, deterministic method of data communications running at 2.5 Mbps. ARCnet passes a token from node to node, and any given node may only communicate when it has the token. ARCnet is widely used in industrial control and BA systems, and has been chosen as a cost-effective LAN topology. The technical specifications of ARCnet are shown in Table 4.3.

ARCnet was intended to be an office automation LAN when it was originally introduced by Datapoint Corporation in the late 1970s. Datapoint proposed a network with distributed computing power operating as one larger computer. This system was referred to as Attached Resource Computer (ARC) and the network that connected these resources was called ARCnet.

The use of ARCnet as an office automation network has diminished and is now dominated by Ethernet. However, ARCnet continues to find success in the industrial and building automation industries because its performance

Table 4.3 ARCnet technical specifications

Specification	Characteristics
Network standard	ANSI 878.1 (close to IEEE 802.4)
Protocol	Standard for layers 1 and 2
Architecture	Bus or star topology
Transmission speed	2.5 Mbps
Network node capacity	254 nodes
Network length	122 m (twisted pair) 305 m (coaxial)
Physical path cabling	Twisted pair or coaxial
Medium Access Control (MAC) method	Token passing

characteristics are well suited for control. ARCnet has proven itself to be very robust. ARCnet also is fast, provides deterministic performance and can span long distances, making it a suitable fieldbus technology.

The term fieldbus is used in the industrial and building automation industries to signify a network consisting of computers, particularly controllers and devices mounted in the 'field' (various locations). ARCnet is an ideal fieldbus. Unlike office automation networks, a fieldbus must deliver messages in a time-predictable fashion. The token-passing protocol of ARCnet provides this timeliness. Fieldbus messages are generally short. ARCnet packet lengths are variable from 0 to 507 bytes with little overhead and are coupled with the high data rate of ARCnet, typically 2.5 Mbps, which is not particularly high these days but is still high enough for field networks in industrial and building automation applications.

4.5.3 LonTalk

LonTalk is a LAN protocol which is closely associated with communication hardware (Neuron chips) also marketed under Echelon Corporation's name. While the BACnet committee elected not to endorse Echelon Corporation's definition of a standard protocol (at higher layers), they did allow for BACnet messages to be sent via LonTalk. The technical specification of LonTalk LANs is shown in Table 4.4.

LonWorks refers to a family of products developed by Echelon Corporation on the basis of the LonTalk protocol. In recent years, several vendors of BA and control equipment have incorporated Echelon technology into some of their products.

All of these products use a proprietary chip called the Neuron chip, which is a dedicated communications processor. Neurons use a technology called LonWorks on the basis of the communications protocol LonTalk for

Table 4.4 LonTalk networks specifications

Specification	Characteristics
Network standard	ANSI/EIA 709.1
Protocol	Standard for layers 1 and 2
Architecture	Bus, star and free topology
Transmission speed	300 bps – 1.25 Mbps
Network node capacity	Nodes per subnet: 127
Network length	130 m (TPT/XF, 1.25 Mbps, 64 nodes) 2700 m (FTT-10, 78 Kbps, bus, 64 nodes) 2200 m (LPT-10, 78 Kbps, bus, 128 nodes)
Physical path cabling	Twisted pair, power line, EIA-485, infrared, optical fibre, coaxial cable, radio-frequency
Media access control (MAC) method	Predictive P-Persistent CSMA

all communications including upper layers. LonMark is the organization responsible for standardization. LonTalk used in LonWorks products has two primary components: lower-layer transport protocol, and an upper-layer application protocol. Here, the term LonTalk refers to the lower transport layers. This distinction is an important one because it directly affects interoperability, and is of relevance to BACnet, which is the focus of Chapter 5.

4.6 Wireless technologies

In recent years, wireless technologies have provided new options that overcome challenging installation constraints and remove physical or financial constraints associated with hard wiring for BA systems. It is widely acknowledged that wireless BA systems can have a great impact on facility performance, cost efficiency and return on investment. Wireless environments are easily adapted to changing business needs or new facility requirements. Also, eliminating wiring and its related effort accelerates the installation process and simplifies retrofitting and system extension. Wireless-based systems offer building owners and facility managers more choices and fewer constraints, including:

- ease of deployment;
- cost benefits;
- scalability of the network;
- simpler and more flexible system design;

- faster and less disruptive installations and retrofits;
- smoother and less costly migrations staged to accommodate budgets and schedules.

Flexibility is the ultimate benefit in deploying a wireless system compared to a wired network, while sensors can be located, or relocated, to optimize system performance, increase customer comfort and adapt to changing floor plans. Many wireless technologies and products are available in the current BAS industry, and a great deal of effort is being expended to develop wireless technologies for BAS applications. Wireless BAS will have great prospects in BAS installations in the very near future, although many aspects still need to be addressed.

There are many short-range wireless communications technologies available now in IT, automation and other mobile data-communication fields. The wireless technologies related to BAS applications are mainly:

- *ZigBee technologies*;
- *802.11 compliant (Wi-Fi) technologies*;
- *Bluetooth technologies*;
- *other proprietary technologies*.

4.6.1 ZigBee technologies

The ZigBee Alliance developed a network specification that is built upon the IEEE 802.15.4 radio. The standard was approved in December 2004 and adds logical network, security and application software. The standard was created to address the market need for a cost-effective, standards-based wireless networking solution that supports low power consumption, low data rates, security and reliability. Potential applications include home automation, BA and automated meter reading, as well as industrial monitoring and control. The range of the ZigBee devices with direct communication is 100 to 300 feet (30–91 m) for typical buildings, which makes it well suited for BA applications.

ZigBee uses multiple network topologies, or configurations – star, mesh or cluster tree (which is a combination of the star and mesh topologies). The most suitable topology varies by facility. Regardless of which topology is used, full function devices (FFDs), such as field panels powered by 24 V AC, can route traffic from other nodes. In contrast, reduced-function devices (RFDs), such as battery-powered, room-temperature sensors, are limited to transmitting and receiving only their own traffic. ZigBee operates at multiple frequencies, 250 Kbps at 2.4 GHz, 40 Kbps at 915 MHz (for US markets) and 20 Kbps at 868 MHz (for European markets). It therefore occupies the same unlicensed band as Wi-Fi and Bluetooth.

4.6.2 Wi-Fi technologies

Wi-Fi technologies are based on the IEEE 802.11 protocol standard (2.4 GHz) which was designed primarily to promote LAN-based product interoperability and is best suited to LAN-based applications. Wi-Fi is primarily used for TCP/IP traffic on Ethernet connections which requires that all devices have a unique IP address. Since wireless BAS field controllers and devices normally need very low data rates, Wi-Fi has no advantage for the applications at field level and therefore is not suitable at this level.

Due to its high speed and being TCP/IP compatible, one potential application of the Wi-Fi standard that does exist for BA is to establish a wireless connection of an Ethernet-ready field panel to a facility's existing wireless LAN.

4.6.3 Bluetooth technologies

Bluetooth is a single-hop point-to-multi-point technology designed for and mainly targeted at ad hoc short-range cable-replacement applications such as wireless keyboards and mobile phone hands-free operation. It can support only a limited number of network devices, up to eight, per network. The data rate and power consumption of Bluetooth radios are less than Wi-Fi and closer to the requirements of BA applications.

4.6.4 Proprietary technologies

There are several proprietary wireless solutions, some of which are well suited for industrial and building automation applications and provide a practical solution for many wireless HVAC I/O applications with data rates ranging from 9,600 bps to over 1 Mbps. Several proprietary wireless network systems emerged on top of the IEEE 802.15.4 physical and MAC layer standard on which ZigBee is also based. Therefore, some proprietary technologies share many of the same technical advantages of ZigBee, including interference-free direct-sequence spread-spectrum transmission and even mesh and mesh-like repeater network protocols. These digital systems provide reliability, performance and useable range, along with advanced transmitters capable of a two- to five-year battery life.

The advantage of proprietary solutions is that they can be optimized to serve relatively small-volume applications. Proprietary solutions do not have the same requirements, and one can optimize them to a higher degree. The disadvantages are higher cost and lack of adequate vendor support for widespread use.

4.6.5 Application of wireless technologies in BAS

Field level: The wireless sensor network needs less bandwidth than the controller network. One can select other lower frequencies to support lower data rates. The devices include sensors and actuators. Power is supplied to wireless sensors by battery, so they must be power-efficient devices. Their capabilities mainly include reporting of value changes and responses to querying. Most of the time they are idle or sleeping to enable power saving. The power of actuators is not an issue as it is usually supplied by AC line power.

Automation level: The wireless controller network needs wider bandwidth to transmit more data than a wireless sensor network. It likely requires the use of a higher frequency band to support higher data rates. The controllers usually are installed where AC line power is available. So power supply is not an issue for the applications at this level. The wireless devices can realize more functions in this level, such as routing messages from one node to another node.

Management level: This level involves computers, network controllers, or other Ethernet-ready devices. The devices at this level usually communicate based on Ethernet. Due to its high traffic and TCP/IP compatibility, Wi-Fi standard can be implemented to establish a wireless connection of an Ethernet-ready field panel to a facility's existing wired or wireless LAN.

Some manufacturers provide wireless BAS products based on Wi-Fi in the management and automation levels. The BAS management level can share a communication network with other information networks, both wired and wireless. Wi-Fi networks are just the infrastructure of the entire building in these cases, and are not specific for BAS in terms of technology and installation. BAS follows the ever-developing technology in the management level in the IT field. Therefore the focus in the BAS field is on the development of wireless networks at automation and field (sensor) levels.

When a large number of wireless sensors need to be networked, several levels of networking may be combined. For example, an IEEE 802.11 (Wi-Fi) mesh network comprised of high-end nodes, such as gateway units, can be overlaid on a ZigBee sensor network to maintain a high level of network performance. A remote application server can also be used in the field close to a localized sensor network to manage the network, to collect localized data, to host web-based applications, and other uses.

Recently in the market, there have been many wireless products applied in BAS, from ICs, modules, sensors, actuators, thermostats, controllers, gateways and transparent transmitters by tunnelling method to total system solutions. The majority of them are compliant with ZigBee or IEEE 802.15.4.

Wireless technologies, particularly ZigBee, are going to have more impact in the BAS industry and wireless building automation is going to be among the practical options. However, they do not yet have a large range of applications. Wireless BAS manufacturers have struggled to make wireless BAS acceptable to more users. A number of issues concern users, including reliability, battery life, security, and interference with other wireless networks or devices.

References

Frost & Sullivan. (2005) *Wireless Sensors in Building Automation*, Frost & Sullivan, Technical Insights Division.

Raimo J. (2005) 'Wireless mesh networks using HVAC controllers', *ASHRAE Journal*, 47(7): 32–6.

—— (2006) 'Wireless mesh controller networks', *ASHRAE Journal*, 48(10): 34–49.

Reed, K. (1996) *Data Network Handbook: an interactive guide to network architecture and operations*, New York: Van Nostrand Reinhold.

Reiss, L. (1987) *Introduction to Local Area Networks with Microcomputer Experiments*, Englewood Cliffs, New Jersey: Prentice-Hall.

Wang, N., Zhang, N. Q. and Wang, M. H. (2006) 'Review wireless sensors in agriculture and the food industry: recent development and future perspective', *Computers and Electronics in Agriculture*, 50: 1–14.

Wang, S. W. (2008) 'Wireless networks and their applications in building automation systems', editorial, *HVAC&R Research*, 14(4): 529–33.

Wills J. (2004) 'Will HVAC control go wireless?', *ASHRAE Journal*, 46(7): 46–52.

5 BAS communication standards

This chapter discusses problems and solutions concerning integration and interoperability of BA systems. Communication standards for BAS networks are the main issue addressed. The most commonly used communication standards in the BA industry, including BACnet, LonWorks, Modbus, PROFIBUS and EIB, are introduced. The standards or common methods of integration at management level are discussed.

5.1 Background and problems

Over the last few decades, incompatibilities and limited opportunities for the integration of BA systems among products of different vendors have frustrated real estate developers, building owners and operators, consultants and system integrators. Although great progress has been made on the interoperability of BA systems, the compatibility problem is still one of the major headaches troubling professionals today. In a typical BAS, usually different communication protocols are employed, even among the products of one company. A popular way to integrate the products of various protocols has been to employ a gateway, which has the role of converting a protocol to another protocol, mapping data points from one protocol to another protocol. But the development of a gateway requires significant effort, and the developer needs to have the technical details of two protocols and understand them as well. It requires great configuration effort to make the gateway map the data points correctly. This makes gateways expensive. The gateway also slows down the response due to the time required for conversion. Furthermore, it is difficult to program and configure a controller through a gateway.

As a practical example, a comprehensive IB teaching laboratory facility that was developed seven years ago in The Hong Kong Polytechnic University is considered here. Figure 5.1 illustrates the configuration of the system, which includes BA system (HVAC control), fire security system, security and access control, lighting control and power monitoring. In this system, all subsystems are integrated into a single Ethernet backbone. Users can supervise and monitor the entire IB system via the management software.

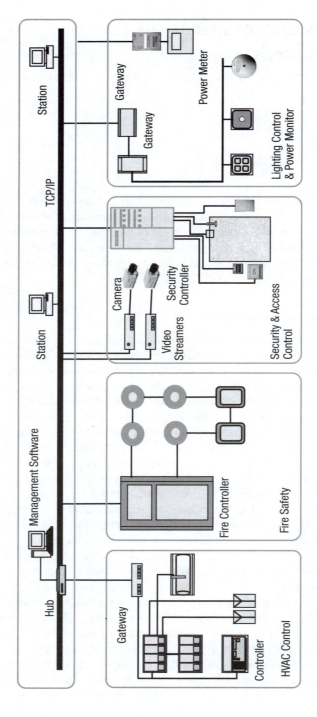

Figure 5.1 An integrated intelligent building system.

Interoperation over different subsystems can be achieved. For example, when a fire alarm occurs, the HVAC system will stop and the relevant image of the space will be displayed. Lighting is turned on automatically when an authorized person intends to enter the laboratory identified by the access control panel. The digital CCTV camera can be turned to monitor designated locations and starts to record when a person intends to access the laboratory or the door is open. The intelligent functions and convenience provided by the integration of the IB subsystems are amazing.

However, every subsystem has a different protocol in this IB system. Gateways are needed to realize conversion of protocols to achieve the total integration of the subsystems. The management software communicates with subsystems via relevant drivers. If interoperation among subsystems is needed, the management software plays the role of 'agent'. This interoperation is a vendor's proprietary method only, not a standard method. It is not flexible and much difficulty will be faced if a new third-party subsystem is to be added to the system.

Nowadays, the rapid development of information technologies offers new possible methods and solutions to overcome these difficulties. To have a clearer understanding of the problem, the hierarchy model of the BAS network is used as a reference. The BAS network can be divided into three levels: management level, automation level and field level. Integration and interoperability can be addressed at different levels. One can achieve integration and interoperation at all three levels starting from the bottom (field level) or achieve integration and interoperation at a higher level. This provides two possible ways to solve the problems of interoperation and integration. One is to employ the same open communication protocol in all the three levels. ISO, ASHRAE and several other organizations have been working on this. For example, ISO adopted a few open protocol standards (e.g. ISO 16484–5 and ISO/IEC 14908–1) from field level to management level to increase the interoperation of BAS. ISO 16484–5 basically refers to *BACnet – A Data Communication Protocol for Building Automation and Control Networks* presented by ASHRAE. The ISO/IEC 14908–1 refers to the LonWorks communication protocol. However, in light of the fact that different protocols are currently in use and the need to integrate BAS and other business systems such as management information systems (MIS), another way is to achieve integration and interoperation with standard protocols at the upper level (e.g. management level) to avoid handling the difference of the lower level protocols directly. For example, OPC (see Section 5.8) and some emerging IT technologies (e.g. XML, SOAP and Web Services) can be employed to solve the problem.

5.2 BACnet and its features

There are a few potential candidates likely in the future to dominate open BAS communication protocols for all the three levels, such as BACnet,

EIB-ObIS (European Installation Bus Object Interface Specification) and LonWorks. They are all object-oriented protocols or have an extension of object-oriented technology. In this section, the issues related to protocol standards are elaborated by addressing mainly BACnet and comparing BACnet with other protocols.

5.2.1 Background of BACnet

BACnet is a data communication protocol for BA and control networks. A data communication protocol is a set of rules governing the exchange of data over a computer network that covers everything from what kind of cable to use to how to form a particular request or command in a standard way. What makes BACnet special is that the rules relate specifically to the needs of BA equipment; that is, they cover things like how to ask for the value of a temperature, define a fan operating schedule or send a pump status alarm.

BACnet has been developed under the auspices of the American Society of Heating, Refrigerating and Air-Conditioning Engineers (ASHRAE). It is an American national standard, a European standard, an ISO global standard and the national standard in more than 30 countries. The protocol is supported and maintained by ASHRAE Standing Standard Project Committee 135. It is the only open protocol that was designed originally for BA from the ground up, and it is an open protocol that supports high-end functions such as scheduling, alarming and trending.

To achieve interoperability across a wide spectrum of equipment, the BACnet specification consists of three major parts. The first part describes a method for representing any type of BA equipment in a standard way. The second part defines messages that can be sent across a computer network to monitor and control such equipment. And the third part defines a set of acceptable LAN architectures that can be used to convey BACnet communications. Let's look at each of these components of the BACnet specification in more detail.

5.2.2 Representing BAS equipment in a standard way – BACnet objects

BACnet provides a standard way of representing the functions of any device, such as analogue and binary inputs and outputs, schedules, control loops, and alarms, by defining collections of related information called 'objects', each of which has a set of 'properties' that further characterize it. Each analogue input, for instance, is represented by a BACnet 'analogue input object' which has a set of standard properties such as present value, sensor type, location, alarm limits and so on. Some of these properties are required while others are optional. One of the object's most important properties is its identifier, a numerical name that allows BACnet to unambiguously access it. Once devices have common 'appearances' on the network in terms of their objects and properties, it is then possible to define messages that can manipulate this information in a standard way. BACnet defines 18 standard object-types at the beginning. As the standard has evolved, object types have been extended from the original 18 to 25 currently. These 25 object types are referred to as standard objects. A BACnet device does not need to support all object types, but if an object type is supported, it must comply with the standard object model for that object type. Each object type has a list of required properties and optional properties. Optional properties can be included at the manufacturer's discretion. These standard objects are listed in Table 5.1.

The purpose of most BACnet objects is clear from the name of the object but some require explanation and are described as follows:

- *Calendar* represents a list of dates that have special meaning when scheduling the operation of mechanical equipment. A list of holidays would be one example.
- *Command* represents a multi-action command procedure, such as a sequenced start-up of several devices.
- *Device* contains general information about a particular device, such as vendor name, model name, location, protocol version supported, object-types supported, etc.
- *Event Enrolment* provides one way to define alarms or other types of events and to indicate who should be notified when they occur. Some objects (Analogue Input, Analogue Output, Analogue Value, Binary Input, Binary Output, Binary Value and Loop) contain optional properties to support intrinsic event-reporting capability and do not need to use Event Enrolment objects.
- *Group* provides a shorthand way to read several values in one request. For example, it may be used to simultaneously update several fields on an operator graphic display.
- *Loop* can be used to represent any feedback control loop, which is some combination of proportional, integral or derivative control.
- *Notification Class* provides a way to manage the distribution of alarm or event notifications that are to be sent to multiple devices.
- *Accumulator* object type defines a standardized object whose properties represent the externally visible characteristics of a device that indicates measurements made by counting pulses. It maintains precise measurement of input count values, accumulated over time.
- *Life Safety Zone* object type defines a standardized object whose properties represent the externally visible characteristics associated with an arbitrary group of BACnet Life Safety Point and Life Safety Zone objects in fire, life safety and security applications.
- *Life Safety Point* object type defines a standardized object whose properties represent the externally visible characteristics associated with initiating and indicating devices in fire, life safety and security applications.

Table 5.1 BACnet standard object types

Analogue Input	Event Enrolment
Analogue Output	File
Analogue Value	Group
Binary Input	Loop
Binary Output	Multi-state Input
Binary Value	Multi-state Output
Calendar	Notification Class
Command	Program
Device	Schedule
Accumulator	Averaging
Trend Log	Multi-state Value
Pulse Converter	Life Safety Point
Life Safety Zone	

5.2.3 Providing standard messages for monitoring and control – BACnet services

BACnet currently defines 40 message types, or 'services', that are divided into six classes. For example, one class contains messages for accessing and manipulating the properties of the objects described in the previous section. A common one is the 'ReadProperty' service request. This message causes the server machine to locate the requested property of the requested object and send its value back to the client. Other classes of services deal with alarms and events, file uploading and downloading, managing the operation of remote devices, virtual terminal functions (accessing equipment across the network as if you were using a directly connected terminal or laptop), and security.

Note that the ability to read and write binary, analogue and text data; schedule control actions; send event and alarm notifications; and carry out similar functions is required by all kinds of BAS equipment, not just HVAC gear. Nonetheless, the committee realized that these capabilities might not cover all situations and developed the standard with an eye towards accommodating future BA and control applications. As a result, one of the real strengths of the BACnet object and services model is that it can be extended easily. If a vendor comes up with some new functionality for which communication is required, the vendor can add new properties to existing object types or create new object types that are accessed in exactly the same way as the 18 defined in the standard. Moreover, a vendor could even dream up new services that go beyond the standard ones. Of course, proprietary features may not be interoperable without vendor cooperation.

Table 5.2 Application layer services

Alarm and event services	Object access services
AcknowledgeAlarm	AddListElement
ConfirmedCOVNotification	RemoveListElement
ConfirmedEventNotification	CreateObject
GetAlarmSummary	DeleteObject
GetEnrolmentSummary	ReadProperty
SubscribeCOV	ReadPropertyConditional
UnconfirmedCOVNotification	ReadPropertyMultiple
SubscribeCOVProperty	ReadRange
UnconfirmedEventNotification	WriteProperty
GetEventInformation	WritePropertyMultiple
LifeSafetyOperation	
Remote device management services	*Virtual terminal services*
DeviceCommunicationControl	VT-Open
ConfirmedPrivateTransfer	VT-Close
UnconfirmedPrivateTransfer	VT-Data
ReinitializeDevice	
ConfirmedTextMessage	*Security services*
UnconfirmedTextMessage	Authenticate
TimeSynchronization	RequestKey
UTCTimeSynchronization	
Who-Has	
I-Have	*File access services*
Who-Is	AtomicReadFile
I-Am	AtomicWriteFile

It is worth noting that while BACnet makes multi-vendor installations possible, it in no way requires the use of multiple suppliers. Since many vendors
will probably choose, and some have already chosen, to use BACnet as their
'native' protocol, you could easily end up with a single-vendor BACnet system. Table 5.2 lists all of the services and shows how they are grouped.

5.2.4 BACnet protocol architecture

As with many other communication protocols, BACnet employs the OSI Model as its reference model. The Open System Interconnection (OSI) Basic Reference Model (ISO 7498) is an international standard that defines a model for developing multi-vendor computer communication protocol standards. The OSI Model addresses the general problem of computer-to-computer communication and breaks this very complex problem into seven smaller, more manageable sub-problems, each of which concerns itself with a specific communication function. Each of these sub-problems forms a 'layer' in the protocol architecture.

However, the OSI Model is just a reference model and does not demand that all layers be realized. BACnet actually implements a collapsed architecture (as shown in Figure 5.2). Only selected layers of the OSI Model are adopted by BACnet to reduce message length and communication processing overhead. Such a collapsed architecture permits the BA industry to take advantage of lower cost, mass-produced processors. As shown in Figure 5.2, BACnet has four layers only, a collapse of the seven-layer architecture.

In Section 5.2, the BACnet object-oriented model and the various message types were discussed. The system designer will still need to select an appropriate network technology to connect everything together. The BACnet committee spent a lot of time on this part of the standard and ended up with six different options as shown in Figure 5.2, each of which fills a particular niche in terms of the price–performance trade-off. The first is Ethernet, the fastest at 10 Mbps and 100 Mbps with 1000 Mbps also recently available. Ethernet is also likely to be the most expensive in terms of cost per device. Next comes ARCnet at 2.5 Mbps. For devices with lower requirements in

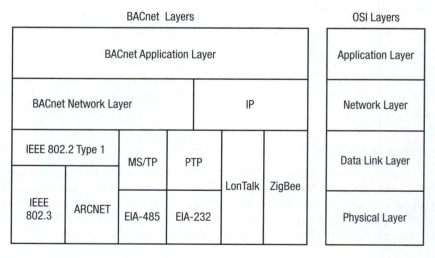

Figure 5.2 BACnet collapsed architecture.

terms of speed, BACnet defines the master–slave/token-passing (MS/TP) network designed to run at speeds of 1 Mbps or less over twisted-pair wiring. Echelon's proprietary LonTalk network can also be used on various media. All of these networks are LANs. BACnet also defines a dial-up or 'point-to-point' protocol called PTP for use over phone lines or hard-wired EIA-232 connections. On 29 January 2009, an addendum to define the use of ZigBee wireless technology as a BACnet data link layer was approved by the American National Standards Institute. A key point is that BACnet messages can, in principle, be transported by any network technology, if and when it becomes desirable to do so.

In fact, it requires more system resources and greater cost to implement all the seven layers in practice. It is also not always a good choice to implement all the layers. Therefore, many protocols, such as the popular TCP/IP (Internet), do not implement all the layers.

In the BA and many other control industries, it is not necessary to implement all seven layers. The four-layer collapsed architecture was chosen after careful consideration of the particular features and requirements of BAS networks, including a constraint that protocol overheads needed to be as small as possible. The use of readily available, widespread technologies, such as Ethernet, ARCnet and LonTalk, will lower costs, increase performance and open new doors for system integration.

5.3 LonWorks and its features

The LonWorks protocol, also known as the LonTalk protocol and the ANSI/ EIA 709.1 Control Networking Standard, is the core of the LonWorks system. The protocol provides a set of communication services that allow the application program in a device to send and receive messages to and from other devices over the network without needing to know the topology of the network or the names, addresses or functions of other devices.

5.3.1 LonWorks protocol architecture

The LonWorks protocol is a layered, packet-based, peer-to-peer communications protocol. It follows the layered architectural guidelines of the ISO Open Systems Interconnection (OSI) Reference Model as shown in Figure 5.3. The Physical Layer is typically implemented using the transceiver from the LonWorks group. There are many options available for different LAN speeds and physical media. By choosing the most suitable transceiver, a LonWorks LAN can be built using most types of LAN cables, including twisted pair, coaxial, power line, radio frequency, infrared and fibre optics. The layers between the Data Link Layer and Presentation Layer are implemented using a Neuron chip.

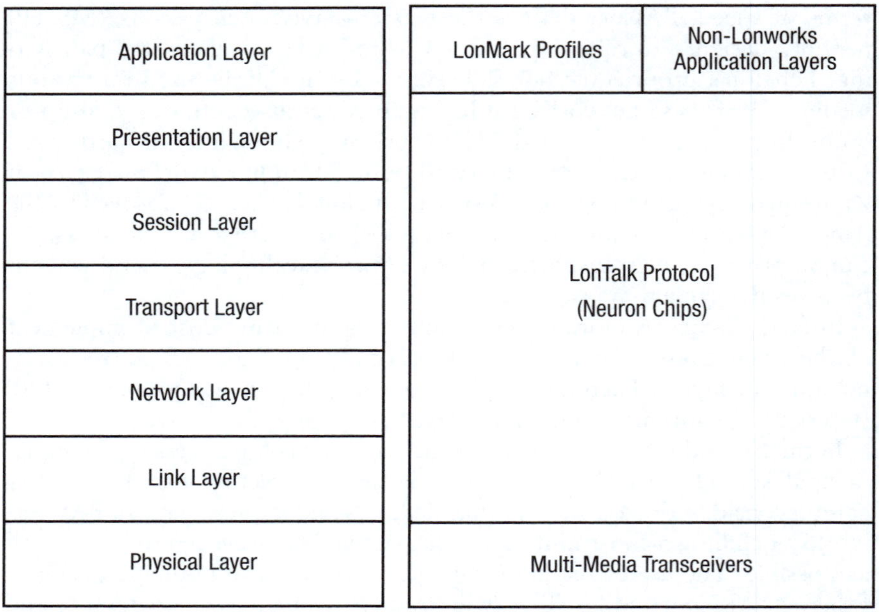

Application Layer	LonMark Profiles	Non-Lonworks Application Layers
Presentation Layer	LonTalk Protocol (Neuron Chips)	
Session Layer		
Transport Layer		
Network Layer		
Link Layer		
Physical Layer	Multi-Media Transceivers	

Figure 5.3 LonWorks protocol architecture.

5.3.2 Addressing

The addressing algorithm defines how packets are routed from a source device to one or more destination devices. Packets can be addressed to a single device, to any group of devices, or to all devices. To support networks of two devices to tens of thousands of devices, the LonWorks protocol supports several types of addresses, from simple physical addresses to addresses that designate collections of many devices. The LonWorks address types include the following:

- *Physical address:* Every LonWorks device includes a unique 48-bit identifier called the Neuron ID. The Neuron ID is typically assigned when a device is manufactured, and does not change during the lifetime of the device.
- *Device address:* A LonWorks device is assigned a device address when it is installed into a particular network. Device addresses are used instead of physical addresses because they support more efficient routing of messages, and they simplify replace failed devices.
- *Group address:* A group is a logical collection of devices within a domain. Unlike a subnet, devices are grouped together without regard for their physical location in the domain.

- *Broadcast address:* A broadcast address identifies all devices with a sub-net, or all devices within a domain.

5.3.3 Message services

The LonWorks protocol offers three basic types of message delivery service and also supports authenticated messages. An optimized network will often use all of these services. These services (listed below) allow trade-offs between reliability, efficiency and security:

- *Acknowledged messaging* provides end-to-end acknowledgement. If acknowledgements are not received, the sender times out and attempts the transaction again. The number of retries and the time-out are both configurable.
- *Repeated messaging* causes a message to be sent to a device or group of any number of devices multiple times. This service is typically used instead of acknowledged messaging because it does not incur the overhead and delay of waiting for acknowledgements.
- *Unacknowledged messaging* causes each message to be sent once to a device or group of any number of devices and no response is expected. This messaging service has the lowest overhead and is the most typically used service.
- *Authenticated service* allows the receivers of a message to determine if the sender is authorized to send that message. Thus, authentication prevents unauthorized access to devices and is implemented by distributing 48-bit keys to the devices at installation time.

LonWorks technology is now a popular LAN technology used in the BA industry, although it is based on proprietary chips. This popularity is due to the fact that although it was born as a proprietary technology of a particular company (later a group of companies), the technology is open and accepted as a standard technology in many cases. The LonTalk/LonWorks LAN capacity well fits the BA network needs at automation and field network levels. It can be implemented on most physical media thanks to the transceivers available. Although the LAN is developed on proprietary chips, those specific networking support chips with supporting software tools provide great convenience for the BA and control system suppliers to develop network technology acceptable by end-users.

5.4 Modbus and its features

Modbus is one of the popular protocols used in industrial automation. Supporting traditional serial protocols of RS232/422/485 and Ethernet protocols allows industrial devices such a PLCs, HMIs and meters to use Modbus as their communication method.

5.4.1 Transmission modes

The Modbus protocol emerged in the mid-1970s as an early protocol for linking terminals with Modicon PLCs using a master–slave (sometimes called a master–client) relationship. It is a simple, open, message-based protocol, and has been accepted as a standard in the industry very quickly. It supports asynchronous point-to-point and multidrop communications and can be used with a variety of serial interfaces (e.g. RS-232, RS-422, RS-485 and modems).

ASCII and RTU are the two possible transmission modes of the original Modbus specifications. Modbus RTU (Remote Terminal Unit) mode is the most common implementation, using binary coding and CRC error checking. Modbus ASCII messages (though somewhat more readable because they use ASCII characters) are less efficient and use less effective longitudinal redundancy check (LRC) error checking. ASCII mode uses ASCII characters to begin and end messages whereas RTU uses time gaps (3.5 character times) of silence for framing. The two modes are incompatible so a device configured for ASCII mode cannot communicate with one using RTU.

Modbus/TCP is a much more recent development, created to allow Modbus ASCII/RTU protocol to be carried over TCP/IP-based networks. Modbus/TCP embeds Modbus messages inside TCP/IP frames. Modbus/TCP sets up connections between nodes on the network, sending requests via TCP in a half-duplex fashion. TCP allows multiple requests to be 'pipelined', or queued in a buffer waiting to be serviced. Modbus/TCP has the capability to use transaction identifiers (sequence numbers).

5.4.2 Architecture and communication process

Modbus devices communicate using a master–slave technique in which only one device (the master) can initiate transactions (called queries). The other devices (slaves) respond by supplying the requested data to the master, or by taking the action requested in the query. A slave is any peripheral device (I/O transducer, valve, network drive, or other measuring device) which processes information and sends its output to the master using Modbus.

Masters can address individual slaves, or can initiate a broadcast message to all slaves. Slaves return a response to all queries addressed to them individually, but do not respond to broadcast queries.

A master's query consists of a slave address (or broadcast address), a function code defining the requested action, any required data, and an error checking field. A slave's response consists of fields confirming the action taken, any data to be returned, and an error-checking field. If no error occurs, the slave's response contains the data requested. If an error occurs in the query received, or if the slave is unable to perform the action requested, the slave will return an exception message as its response. The error-checking field of the message frame allows the master to confirm that the contents of the message are valid. Additionally, parity checking is also applied to each transmitted character in its data frame.

5.4.3 Message structure

The Modbus application protocol uses a well-defined message format. Each Modbus message has the same structure involving four basic elements, including *device address* (address of receiving device), *function code* (Modbus function code), *data* (data block with additional information), and *error checking*. The sequence is always the same, which makes passing very fast and efficient.

5.5 PROFIBUS and its features

PROFIBUS (PROcess Fieldbus) is a widely accepted open automation and field network standard, which is supported by an industry supplying a wide range of equipment, tools and support. PROFIBUS was introduced in 1989 as a German standard, DIN 19245, later adopted as International Standard EN 50170. The PROFIBUS standard is now incorporated into IEC 61158, the international fieldbus standard.

5.5.1 Protocol mode and technologies

The PROFIBUS family consists of three compatible versions offering very high integrity and a capability appropriate to the need, including:

- *PROFIBUS DP (Decentralized Periphery)* providing low-cost, high-speed, simple field-level communications. About 90 per cent of current PROFIBUS applications use DP.
- *PROFIBUS FMS (Fieldbus Message Specification)* provides high-end, applications-level communications. It is normally used at cell or controller level, providing object-oriented transmission of structured data, loading and control of programs, alarm services, and the like.
- *PROFIBUS PA (Process Automation)* has been developed specifically for very cost-effective two-wire connection carrying both power and data. It is particularly cost effective for hazardous environments such as in industrial processing.

The three protocol technologies can operate together. DP and FMS share the same electrical transmission system (RS-485), while PA uses a different electrical transmission system (IEC 1158–2) but shares the same protocol as DP and FMS. PROFIBUS DP extensions and the integration of PROFIBUS with Ethernet technology indicate that FMS is less important than in the past. FMS is no longer supported by PROFIBUS International. However, there are still FMS installations in operation.

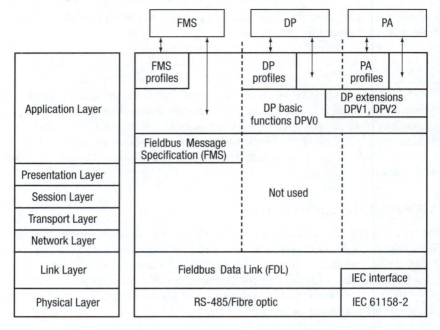

Figure 5.4 PROFIBUS protocol model.

Besides BA, PROFIBUS has attracted applications across a wide range of industries, including factory automation, robotics and machine tools, food production, and chemical and petrochemical production.

All three PROFIBUS versions (DP, FMS and PA) use a uniform bus access protocol. This protocol is implemented by layer 2 of the OSI Reference Model as shown in Figure 5.4. In PROFIBUS, layer 2 is called the *Fieldbus Data Link* (FDL). The FDL handles transmission protocols and includes data security and error-detection measures. PROFIBUS can be implemented over three optional transmission systems (cables), each with its own area of applicability. These include: RS485, optical and shielded twisted pair.

5.5.2 Architecture and communication process

Data is exchanged in PROFIBUS systems using *messages* or *telegrams* that pass between stations. A PROFIBUS network consists of several stations, including *masters* or *slaves*. Master stations (also called *active stations*) control the bus communication. Slave stations (also called *passive stations*) can only respond to a request from a master. There are two types of master station: Class 1 and Class 2. The Class 1 masters include PLCs, controllers, SCADA stations, etc. The Class 2 masters include configuration tools, bus monitors and diagnostics. Slaves include I/O blocks, transmitters, actuators, valves and drives.

PROFIBUS systems can have one or more masters and many slaves. Each master can control (communicate with) one or more slaves. All stations have the same priority. No master is more important than any other. No slave is more important than any other. In single master systems, the master cyclically talks to each slave in turn. In multi-master systems the masters pass a 'token' around, giving rights to control the bus. When a master has the token it can communicate with its slaves. PROFIBUS networks can have different optional speeds. Standard PROFIBUS data rate options include: 9.6, 19.2, 45.45, 93.75, 187.5, 500 Kbps, and 1.5, 3.0, 6.0, 12.0 Mbps.

5.6 EIB and its features

The European Installation Bus (EIB) is a fieldbus designed to enhance electrical installations in homes and buildings of all sizes by separating the transmission of control information from the traditional main wiring. EIB is based on an open specification maintained until recently by the EIB Association (EIBA). The key parts of it were included in CEN1998 and EIA1999. In 2002, EIB was merged with Batibus and European Home System (EHS). The target of this merger was to create a single European Home and Building Electronic System standard. The new KNX standard (Konnex2004) seeks to combine their best aspects.

EIB/KNX already provided the choice of dedicated twisted-pair cabling and power line transmission as well as a simple form of IP tunnelling. RF communication and advanced IP tunnelling were added under the KNX umbrella.

5.6.1 EIB/KNX

The main EIB/KNX medium is the twisted-pair cabling variant now known as KNX TP1. The single twisted pair carries the signal as well as 29 V DC to power devices with up to 50 mW from Class 2 (Safety Extra Low Voltage) power supply. Data is transmitted using a balanced base band signal with 9,600 bps. TP1 allows free topology wiring with up to 1000 m cable length per physical segment. Up to four segments can be concatenated using bridges (called line repeaters), forming a line. Medium access on TP1 is controlled using CSMA with bit-wise arbitration on message priority and station address. Four priority levels are provided.

KNX RF uses a sub-band in the 868 MHz frequency band reserved for short-range devices by European regulatory bodies which is limited by a duty cycle requirement of less than 1 per cent. Particular attention was given to minimizing hardware requirements. To this end, KNX RF not only supports bidirectional communication but transmit-only devices as well. This reduces the cost of simple sensors and switches without status indicators. KNX RF devices communicate in a peer-to-peer method.

5.6.2 EIBnet/IP

EIBnet/IP addresses tunnelling over IP networks. Its core framework supports discovery and self-description of EIBnet/IP devices. It currently accommodates the specialized 'Service Protocols' Tunnelling and Routing. Tunnelling enables point-to-point communication between two EIBnet/IP devices. Its primary application focus is to provide remote maintenance access to EIB/KNX installations in an easy-to-use manner. Routing allows the use of an IP backbone to connect multiple EIB/KNX sub-installations. EIBnet/IP routing is designed to work 'out of the box' as far as possible. Routers communicate using UDP multicast. Group management relies on Internet Group Management Protocol (IGMP) and no central configuration server is needed. The basic building block of an EIB network holds up to 254 devices in free topology. Following a three-level tree structure, subnets are integrated by connecting to the main network via routers. IP tunnelling is typically used for the main network and the backbone, with EIBnet/IP routers linking sub-networks. Overall, the network can contain roughly 60,000 devices at maximum.

5.7 Compatibility of different open protocol standards

5.7.1 Can a BACnet device interoperate with a LonWorks device?

One might think that BACnet devices should be able to interoperate with LonWorks devices as BACnet includes LonTalk. It is not the case in reality.

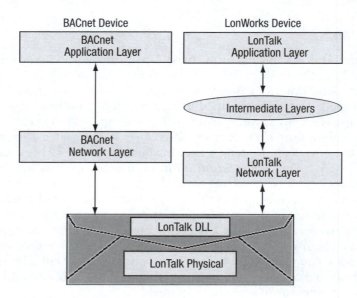

Figure 5.5 BACnet transportation on LonTalk.

BACnet adopts only the subset of LonTalk, not all layers. Only the LonTalk Physical Layer and LonTalk Data Link Layer are adopted as one option of its six alternative transportation technologies. The other layers of BACnet and LonWorks are completely different. Therefore, BACnet devices and LonWorks devices are not interoperable.

The reason for the lack of interoperability is not difficult to understand as can be seen in Figure 5.5. BACnet only adopts the LonTalk Physical Layer and Data Link Layer as its transportation mechanism. The upper layers are completely different. It is similar to the way that letters written in different languages can be put into envelopes; the envelopes can be the same but the letters are in different languages. Only BACnet devices can read BACnet messages. Similarly, only LonWorks devices can read LonWorks messages.

5.7.2 IP compatibility of BACnet

BACnet devices are easy to connect to an intranet or the Internet. They have various choices of devices, such as BACnet router, BBMD, BACnet PAD and even BACnet devices themselves (native Ethernet 802.3 devices, native BACnet/IP devices).

There are two kinds of technologies for BACnet to realize an IP connection. One is BACnet/IP Packet-Assembler-Disassembler (PAD) router technology, which is specified in Annex H.3 of ANSI/ASHRAE Standard 135–2001. A PAD router is placed on every BACnet network that is to be connected over an IP network to another BACnet network. When it receives a BACnet message for a device on another BACnet network which is reachable only through an IP inter-network, it puts the BACnet message into a UDP/IP message with the IP address of the PAD device on the destination BACnet network and sends the message over the IP network. The receiving PAD extracts the BACnet message and transmits the message on the local network. Since the PAD sends the message directly to a known IP of another PAD, this requires that the PAD maintain a table of its peer PAD devices. Unfortunately, tunnel routers have several significant configuration issues and require a lot of expert manual configuration (maintaining the tables) to work correctly. Another problem is that the addition of a single BACnet device to a BACnet system via an IP inter-network would require either that the device itself provides the PAD service or the addition of another device that provides the PAD service. This would significantly increase the cost.

BACnet has improved significantly, overcoming these limitations by adopting the second technology for BACnet to realize an IP connection; that is, BACnet/IP defined in Annex J of ANSI/ASHRAE Standard 135–2001. Basically, BACnet/IP devices communicate using IP messages in lieu of BACnet messages, allowing the devices to be added easily to any IP inter-network. This is implemented by defining a new protocol layer called the BACnet Virtual Link Layer (BVLL). Broadcast management is accomplished by defining the capabilities of a new device called a BACnet Broadcast

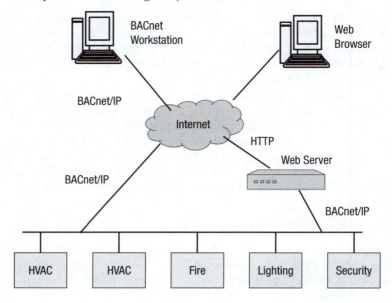

Figure 5.6 Connection of BACnet and Internet.

Management Device (BBMD). Alternatively, IP multicast may be used. This method allows devices to enter the system from anywhere in the IP inter-network. It supports 'native IP' BACnet devices such as unitary controllers which send and receive BACnet messages within IP frames. Therefore, BACnet devices effectively use IP inter-networks as BACnet LANs. Figure 5.6 shows the connection between BACnet and Internet and how to access BACnet devices with a web browser.

5.8 Integration at management level

Integration at management level is basically the communication of applications. A few technologies can be used for this purpose. Two decades ago, the majority of products adopted proprietary communication protocols while Microsoft DDE technology was used to achieve communication of different software. However, due to the poor performance of DDE, this technology has been phased out. Various communication technologies have been developed to achieve suitable communication among applications, such as OPC (OLE for Process Control), COM/DCOM, CORBA and Web Services. These IT technologies, particularly OPC and Web Services technology, have a great influence on and are being adopted in the BAS communication field.

Integration technologies at management level should not only allow the integration of different BASs, but should also allow the integration of BAS and other enterprise applications (e.g. Enterprise Resource Planning – ERP).

5.8.1 OPC technology

Based on Microsoft's OLE, COM (Component Object Model) and DCOM (Distributed Component Object Model) technologies, traditional OPC (i.e. OPC DCOM) consists of standard interfaces, properties and methods for use in process control and manufacturing automation applications.

One of the valuable features of OPC is that it provides a common interface for communicating with diverse process-control devices, regardless of the controlling software or protocols used in the process. Before OPC became available, application developers had to create specific communications drivers for each control system with which they wanted to connect. With OPC, application vendors no longer need separate drivers for each new processor or protocol. Instead, manufacturers create a single optimized OPC driver for their product, as illustrated in Figure 5.7.

OPC can play a vital role in the integration of different vendors' BA

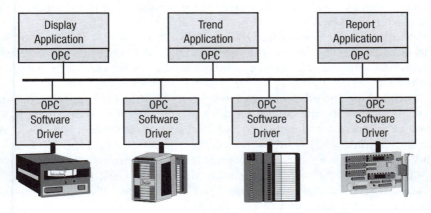

Figure 5.7 Communication of automation systems using OPC.

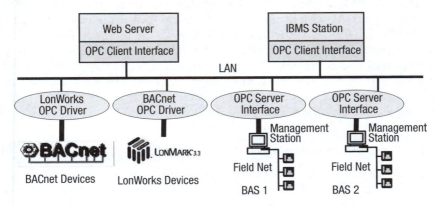

Figure 5.8 BAS devices/systems integration via OPC.

systems or devices. For example, in Figure 5.8, the management software platforms of BAS 1 and BAS 2 have provided the OPC server interface, so Intelligent Building Management System (IBMS) station software and web applications can communicate with them via their OPC interfaces. It is easier if the OPC server driver provided is accompanied with a device that can be connected directly via the OPC server driver. However, there are still problems with OPC to be solved, as discussed below.

OPC DCOM uses COM and DCOM as the core technology for the software interface. Therefore, when you use an OPC server located on a machine different from your own, you must configure the DCOM security. Many installers experienced this as a problem particularly on Windows 95 and Windows 98 machines as there is a security infrastructure on those versions of Windows, different from Windows NT/2000/XP. As a result, DCOM security is often disabled, which leads to severe security risks. It is even riskier to use an OPC server over the Internet. As OPC DCOM interaction over the Internet will result in severe security problems, it is not practical to be used in this way.

Another problem of using OPC DCOM servers is that they cannot be accessed from a non-Windows system, so OPC DCOM clients cannot communicate with non-Windows systems. Therefore, they cannot be used on other platforms.

For tracking and other monitoring software in BAS, it is important that all data is received by the application without interruption. When an application experiences a bad connection or even gets disconnected, OPC DCOM cannot automatically try to reconnect. This is another problem of OPC in BAS applications which needs to be addressed in its further development.

A new OPC XML-DA standard was developed taking into account the disadvantages discussed above. With this new standard, OPC allowed manufacturing and process data to be accessed via the Internet for the first time. In this case, the OPC server is configured as a Web Service.

The OPC Foundation developed the next generation of OPC, OPC Unified Architecture (UA), with the first parts of the specification released in August 2006. The full OPC UA specification is still a work in progress. There will be no commercially usable UA products before all parts of the UA specification are released and compliance test tools are available. For applications that need only Data Access (OPC DA) functionality XML DA is a viable alternative. XML DA has the same communication security as UA and DCOM is not used, eliminating the frustrating DCOM issues.

5.8.2 Web Services technology

Web Services is a new and powerful model for creating applications from reusable software models supported on the Internet or control networks using HTTP technology, which provides loosely coupled, flexible and dynamic solutions by using emerging techniques such as SOAP, WSDL and

UDDI. These technologies are designed on the basis of existing web technologies. Web Services are self-contained, self-describing, modular applications that can be published, located and invoked across the Web. The core technologies of Web Services are as follows:

1 *XML:* XML is developed first, and a Web Service is developed on the basis of XML. All Web Service definitions and message are in the form of XML.
2 *Universal Description, Discovery and Integration (UDDI):* UDDI provides the protocol to register and find a Web Service in a public UDDI directory.
3 *Web Service Definition Language (WSDL):* WSDL is an XML Document Type Definition (DTD) that defines the standard format of describing a Web Service. Documenting a Web Service in WSDL provides plenty of scope for future automatic client site Web Service stub generation or using graphical user interface (GUI) applications to configure or plug in a Web Service.
4 *Simple Object Access Protocol (SOAP):* SOAP is also an XML DTD that defines how a message is sent over the line. As long as you send a service request in SOAP, no matter how the Web Service is constructed, it will serve you. SOAP offers vendor, platform and language independence. With SOAP, developers can easily bridge applications written with COM, CORBA or Enterprise JavaBeans.

Web Services adopts common Internet protocols (e.g. HTTP, HTTPS and SMTP) as its basic communication framework. These protocols' ports usually have a firewall, so Web Services are Internet-friendly. Using the Web Service technology, BAS systems from different vendors, even on different platforms, can be integrated easily, as illustrated in Figure 5.9.

As shown in Figure 5.9, the portal application can access different BAS

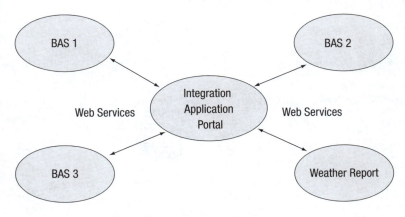

Figure 5.9 BAS systems integration using Web Services.

via Web Services and the non-BAS systems can be easily integrated as well. For example, a weather bureau could offer a Web Service that allows a BA system to automatically retrieve temperature forecast data for use by various control algorithms. Similarly, the BA system itself could offer a Web Service that allows a tenant's accounting system to obtain up-to-the-minute figures on energy consumption.

However, since the SOAP request/response is enveloped in XML format, it is too complex for control communications in some situations and increases the need for processor power and additional response time.

5.8.3 Application issues

While traditional distributed communication technology platforms, including DCOM, CORBA and JAVA/RMI, provide great implementation vehicles for integration, none of them is a clear winner. The strengths of Web Services as an information integrator and distributor, namely simplicity of access and ubiquity, are important in resolving the fragmented middleware world where interoperability is hard to come by. Web Services complements these platforms by providing a uniform and widely accessible interface and access glue over services that are more efficiently implemented in a traditional middleware platform. The combination of OPC DCOM and Web Services is a good example.

On an intranet, OPC DCOM configuration and security can be addressed relatively easily in comparison with the situation on the Internet. OPC DCOM can achieve better communication performance on intranet compared with Web Services. Therefore, OPC DCOM is a better choice on intranet. However, when it is extended to the Internet, Web Services can provide better security and integration performance compared with OPC DCOM. In conclusion, the best way is to combine OPC DCOM and Web Services in application when the TCP/IP-based control network is connected to the Internet.

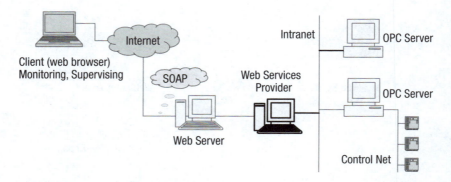

Figure 5.10 BAS integration combining the use of OPC DCOM and Web Services.

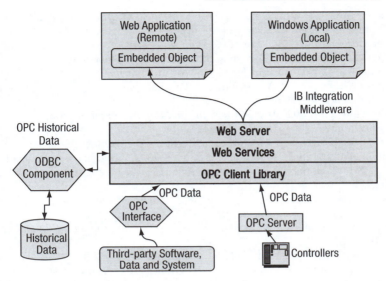

Figure 5.11 IB integration middleware based on OPC and Web Services technologies.

Figure 5.10 shows a schematic of an IB system integration and management tool developed in the IB laboratory of The Hong Kong Polytechnic University. OPC DCOM is employed in the TCP/IP LAN network, integrating the control networks at a lower level. The OPC client components are wrapped into the campus network (or public) as Web Services. Users can access the Web Services to supervise and control the BAS system via campus intranet/Internet. Figure 5.11 illustrates the IB integration middleware architecture of the tool based on OPC DCOM and Web Services technologies used to realize integration and interoperability among BAS systems.

References

ANSI/ASHRAE Standard. (2001) *BACnet® – a data communication protocol for building automation and control networks*, Atlanta, Georgia: American Society of Heating, Refrigerating, and Air-Conditioning Engineers.

—— (2007) *BACnet® – a data communication protocol for building automation and control networks*, Atlanta, Georgia: American Society of Heating, Refrigerating, and Air-Conditioning Engineers.

BACnet Manufacturers Association. (2000) 'Interoperability: going beyond the standard', *Maintenance Solutions*, September, pp. 16–17.

Bushby, S. T. (1998) 'Communication gateways: friend or foe?', *ASHRAE Journal*, 40(4): 50–3.

Bushby, S. T. and Newman, H. M. (2002) 'BACnet today', *ASHRAE Journal* (Suppl.), 44(10): S10–18.

Chisholm, A. (1998) *OPC Data Access 2.0 Technical Overview*. Available at: www. opcfoundation.org/04_ tech/opcae-short.ppt

Echelon Corporation. (2002) *LonMark Application-Layer Interoperability Guidelines – Version 3.3*, San Jose, California: Interoperability Association.

EIA Standard. (2002) *Control Network Protocol Specification*, EIA/CEA-709.1-B (Revision of EIA-709.1-A), Washington, DC: Energy Information Administration, US Department of Energy.

Frost & Sullivan. (2002) *North American Building Automation Protocol Analysis*, Frost & Sullivan Report A143–19, May.

Kranz, H. R. (2001) 'Standard protocols: what is their influence on the world of building automation?', Proceedings of Clima2000, Napoli, Italy, September.

Li, T. (2001) 'Web Services technology'. Available at: www.acssnl.org/acssnlnews/ liting.pdf

Lu, W. L. (2002) 'Asia situation and development of intelligent building', *Engineering Design CAD and Intelligent Building*, 67(6): pp. 34–6.

Piper J. (2001) 'Understanding open protocols', *Building Operating Management*, 48(8): 42–5.

Wang, S. W., Xu, Z. Y., Li, H., Hong, J. and Shi, W. Z. (2004) 'Investigation on intelligent building standard communication protocols and application of IT technologies, automation in construction', *Automation in Construction*, 13(5): 607–19.

6 Internet technologies and their applications in BASs

The Internet (the short form of internetwork) is now a very popular tool and Internet technologies are widely used in industry and daily life. Due to the Internet's popularity, market pressure has led to Internet technologies being open, standard and low cost. Thus, these technologies are gaining more and more influence and applications in the BAS and IB industry. This chapter provides a background to the Internet, basic information on Internet protocols, a comparison between BAS LAN and the Internet network, information on typical Internet technologies, an outline of the influence and applications of Internet technologies at automation level and management level in the BAS/IB field, and discussion of the total integration of IB systems.

6.1 Background of the Internet

6.1.1 What is an internet?

An *internet* (the shortened form of internetwork) is a collection of individual networks, connected by intermediate networking devices, that functions as a single large network. Internetworking refers to the industry, products and procedures that meet the challenge of creating and administering an internet. Figure 6.1 illustrates some different kinds of network technologies that can be interconnected by routers and other network devices to create an internet.

The most well-known example of internetworking is the Internet, a networking of networks based on many underlying hardware technologies, but unified by an internetworking protocol standard, the Internet Protocol Suite (TCP/IP).

6.1.2 History of the Internet

In 1957, the US government formed the Advanced Research Projects Agency (ARPA), a section of the Department of Defense charged with ensuring US leadership in science and technology with military applications. In 1969, ARPA established ARPANET, the forerunner of the Internet.

ARPANET was a network that connected major computers at the University

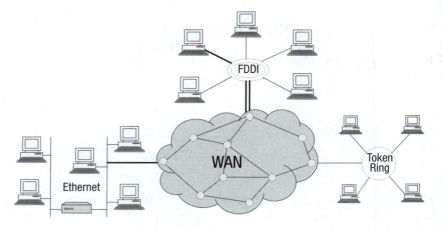

Figure 6.1 Different network technologies connected to create an internet.

of California at Los Angeles, the University of California at Santa Barbara, Stanford Research Institute, and the University of Utah. Within a couple of years, several other educational and research institutions joined the network.

In response to the threat of nuclear attack, ARPANET was designed to allow continued communication if one or more sites was destroyed. Unlike today, when millions of people have access to the Internet from home, work or their public library, ARPANET served only computer professionals, engineers and scientists.

Throughout the 1970s, developers created the protocols used to transfer information over the Internet. By the early 1980s, Usenet newsgroups and electronic mail had been launched. Most users were affiliated with universities, although libraries began to connect their catalogues to the Internet as well. During the late 1980s, developers created indices, such as Archie and the Wide Area Information Server (WAIS), to keep track of the information on the Internet. In 1989, the World Wide Web came into being thanks to developer Tim Berners-Lee and others at the European Laboratory for Particle Physics, also known as CERN. To give users a friendly, easy-to-use interface to work with, the University of Minnesota created its Gopher, a simple menu system for accessing files, in 1991.

6.2 Internet protocols

The Internet protocols are the world's most popular open-system (non-proprietary) protocol suite because they can be used to communicate across any set of interconnected networks. They are equally well suited for LAN and WAN communications. The Internet protocols consist of a suite of communication protocols, of which the two best known are the Transmission

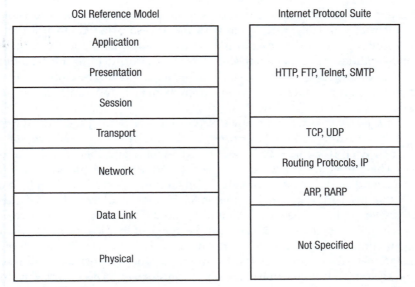

Figure 6.2 Internet protocols span the complete range of OSI model layers.

Control Protocol (TCP) and the Internet Protocol (IP). The Internet protocol suite not only includes lower-layer protocols (such as TCP and IP), but it also specifies common applications such as electronic mail, terminal emulation and file transfer, etc.

Documentation of the Internet protocols (including new or revised protocols) and policies are specified in technical reports called 'Request for Comments' (RFCs), which are published and then reviewed and analyzed by the Internet community. Protocol refinements are published in the new RFCs. To illustrate the scope of the Internet protocols, Figure 6.2 maps typical protocols of the Internet protocol suite and their corresponding OSI model layers. Internet protocols at Network Layer and Transport Layer are discussed in more detail in this chapter.

Address Resolution Protocol (ARP) is a network protocol which maps a Network Layer protocol address to a Data Link Layer hardware address. For example, ARP is used to resolve an IP address to the corresponding Ethernet address. ARP is extensively used by all the hosts in an Ethernet network. Reverse Address Resolution Protocol (RARP) is a TCP/IP protocol that permits a physical address, such as an Ethernet address, to be translated into an IP address.

6.2.1 Internet Protocol (IP)

The Internet Protocol is a Network Layer protocol that contains addressing information and some control information that enables packets to be routed.

IP is documented in RFC 791 and is the primary Network Layer protocol in the Internet protocol suite. Along with the Transmission Control Protocol (TCP), IP represents the heart of the Internet protocols. IP has two primary responsibilities: providing connectionless, best-effort delivery of datagrams through an internet, and providing fragmentation and reassembly of datagrams to support data links with different maximum-transmission unit sizes.

IP addressing: As with any other Network Layer protocol, the IP addressing scheme is integral to the process of routing IP datagrams through an internet. Each IP address has specific components and follows a basic format. These IP addresses can be subdivided and used to create addresses for sub-networks, as discussed in more detail later in this section. Each host on a TCP/IP network is assigned a unique 32-bit logical address that is divided into two main parts: the network number and the host number. The network number identifies a network and must be assigned by the Internet Network Information Center (InterNIC) if the network is to be part of the Internet. An Internet service provider (ISP) can obtain blocks of network addresses from the InterNIC and can itself assign address space as necessary. The host number identifies a host on a network and is assigned by the local network administrator.

IP address format: The 32-bit IP address is grouped 8 bits at a time, separated by full stops ('dots'), and represented in decimal format (known as *dotted decimal notation*). Each bit in the octet has a binary weight (128, 64, 32, 16, 8, 4, 2, 1). The minimum value for an octet is 0, and the maximum value for an octet is 255. Figure 6.3 illustrates the basic format of an IP address.

IP address classes: IP addressing supports five different address classes: A, B, C, D and E. Only classes A, B, and C are available for commercial use. The left-most (high-order) bits indicate the network class. The high-order bits in each class are fixed. Figure 6.4 illustrates the format of the commercial IP address classes.

The class of address can be determined easily by examining the first octet of the address and mapping that value to a class range in the following table. In

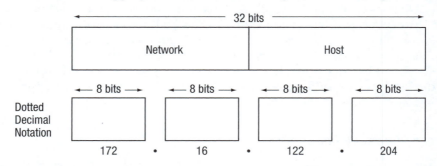

Figure 6.3 An IP address consists of 32 bits, grouped into four octets.

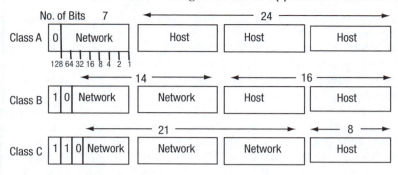

Figure 6.4 IP address formats A, B, and C for commercial use.

an IP address of 172.31.1.2, for example, the first octet is 172. Because 172 falls between 128 and 191, 172.31.1.2 is a Class B address. Table 6.1 summarizes the range of possible values for the first octet of each address class.

IP subnet addressing: IP networks can be divided into smaller networks called sub-networks (or subnets). Subnetting provides the network administrator with several benefits, including extra flexibility, more efficient use of network addresses and the capability to constrain broadcast traffic (a broadcast will not cross a router). Subnets are under local administration. As such, the outside world sees an organization as a single network and has no detailed knowledge of the organization's internal structure. A given network address can be broken up into many sub-networks. For example, 172.16.1.0, 172.16.2.0, 172.16.3.0, and 172.16.4.0 are all subnets within network 172.16.0.0. (All 0s in the host portion of an address specify the entire network.)

6.2.2 TCP and UDP

Transmission Control Protocol (TCP): The TCP provides reliable transmission of data in an IP environment. TCP corresponds to the Transport Layer of the OSI Reference Model. Among the services TCP provides are stream

Table 6.1 The range of possible values existing for the first octet of each address class

Address class	First octet in decimal	High-order bits
Class A	1–126	0
Class B	128–191	10
Class C	192–223	110
Class D	224–239	1110
Class E	240–254	1111

data transfer, reliability, efficient flow control, full-duplex operation and multiplexing.

With stream data transfer, TCP delivers an unstructured stream of bytes identified by sequence numbers. This service benefits applications because they do not have to chop data into blocks before handing it off to TCP. Instead, TCP groups bytes into segments and passes them to IP for delivery.

TCP offers reliability by providing connection-oriented, end-to-end reliable packet delivery through an internet. Bytes not acknowledged within a specified time period are retransmitted. The reliability mechanism of TCP allows devices to deal with lost, delayed, duplicate or misread packets. A time-out mechanism allows devices to detect lost packets and request retransmission.

User Datagram Protocol (UDP): The UDP is a connectionless Transport Layer protocol that belongs to the Internet protocol family. UDP is basically an interface between IP and upper-layer processes. UDP ports distinguish multiple applications running on a single device from one another.

Unlike the TCP, UDP adds no reliability, flow-control or error-recovery functions to IP. Because of UDP's simplicity, UDP headers contain fewer bytes and consume less network overhead than TCP. UDP is useful in situations where the reliability mechanisms of TCP are not necessary, such as in cases where a higher-layer protocol might provide error and flow control.

UDP is the transport protocol for several well-known Application Layer protocols, including Network File System (NFS), Simple Network Management Protocol (SNMP), Domain Name System (DNS), and Trivial File Transfer Protocol (TFTP). Domain Name System (DNS) is an Internet service that translates domain names (meaningful names to humans) into IP numbers. For example, when a user uses 'xxx.polyu.edu.hk', it will be addressed to the DNS server of PolyU for translation into the corresponding IP number.

6.2.3 Internet Application Layer protocols

The Internet Protocol suite includes many Application Layer protocols that represent a wide variety of applications, including the following. Table 6.2 lists some typical application layer protocols and the applications that they support.

- *File Transfer Protocol (FTP):* Moves files between devices.
- *Hypertext Transfer Protocol (HTTP):* Transmits hypertext over networks. This is the protocol of the World Wide Web.
- *Simple Network-Management Protocol (SNMP):* Primarily reports anomalous network conditions and sets network threshold values.
- *Telnet:* Serves as a terminal emulation protocol.
- *X Windows:* Serves as a distributed windowing and graphics system used for communication between X terminals and UNIX workstations.

Table 6.2 Some application layer protocols and their applications

Application	Protocols
File transfer	FTP
Terminal emulation	Telnet
Electronic mail	SMTP
Network management	SNMP
World Wide Web	HTTP
Distributed file services	NFS, XDR, RPC, X Windows

- *Network File System (NFS), External Data Representation (XDR), and Remote Procedure Call (RPC):* Work together to enable transparent access to remote network resources.
- *Simple Mail Transfer Protocol (SMTP):* Provides electronic mail services.
- *Domain Name System (DNS):* Translates the names of network nodes into network addresses.

6.3 Internet LAN vs WAN

6.3.1 A comparison between LAN and WAN

Local area networks (LANs) evolved following the revolution of PCs, as discussed in Chapter 3. LANs enabled multiple users in a relatively small geographical area to exchange files and messages, as well as to access shared resources such as file servers and printers.

A wide area network is a data communications network that covers a

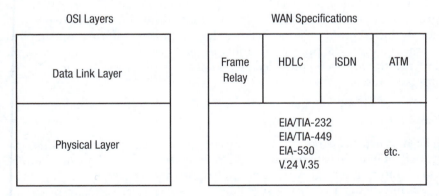

Figure 6.5 Examples of WAN technologies at the two lowest layers of the OSI Model.

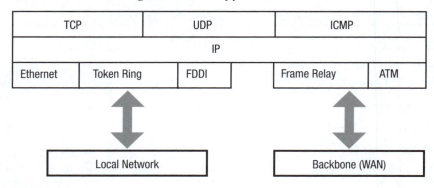

Figure 6.6 IP on LAN and WAN.

relatively broad geographic area and that often uses transmission facilities provided by common carriers, such as telecommunication companies. WAN technologies generally specify the lower layers of the OSI Reference Model, i.e. Physical Layer, Data Link Layer and sometimes Network Layer. Figure 6.5 illustrates some examples of WAN protocols normally used at the two lower layers of the OSI Model. Asynchronous Transfer Protocol (ATM), Integrated Services Digital Network (ISDN), Frame Relay and High Level Data Link Control (HDLC) are popular Data Link Layer protocols of WAN.

WANs interconnect LANs with geographically dispersed users to create connectivity. Today, high-speed LANs and Internet are becoming widely used, largely because they operate at very high speeds and support such high-bandwidth applications as multimedia and video conferencing.

IP over WAN has been developing very quickly in recent years. Figure 6.6 illustrates the use of IP and some related protocols in LAN and Internet (WAN) backbone supported by other protocols at upper and lower layers.

The Internet could be considered the largest WAN ever. However, it would be a mistake to consider the Internet as just a computer network, or even a group of networks connected to one another. These computer networks are simply the medium or infrastructure that carries the information supplied by the Internet. The beauty and utility of the Internet lie in the information, and therefore services, provided as well as the connectivity provided by the infrastructure.

6.3.2 Intranet

The Internet has captured world attention in recent years. In reality, the growth of internal networks based on Internet technologies, known as intranets, is outpacing the growth of the global Internet itself.

An intranet is a company-specific network that uses software programs based on the Internet TCP/IP protocol and common Internet user interfaces such as the web browser. In simple terms, an intranet is the application of

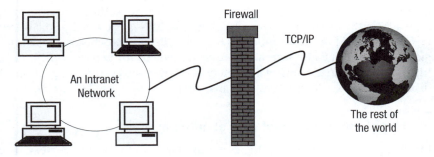

Figure 6.7 An intranet: a company-wide network based on Internet technologies.

Internet technologies within a private LAN or WAN network, usually of an organization.

An intranet uses TCP/IP, HTTP and other Internet protocols. In general, it looks like a private version of the Internet. The intranet environment is completely owned by an enterprise and is generally not accessible from the Internet at large. Today, many intranets are built around web servers delivering HTML pages. Private messages can be transmitted safely from one party to another within private networks.

Typically, larger enterprises allow users within their intranet to access the public Internet through firewall servers that have the ability to screen messages in both directions so that company security is maintained. Companies can also send private messages through the public network, with special encryption/decryption and other security safeguards to connect one part of their intranet to another.

6.4 An overview of applications of Internet technologies in BAS

Both BAS networks and Internet can be data (including multimedia messages in data format) communication networks. Their basic principles are the same, which means one can adopt the technologies of the other conveniently. In fact, along with the rapid development of information technologies, BAS has adopted many popularly used information technologies, such as Internet and intranet technologies. From BAS protocols to system integration technologies, Internet and intranet technologies have played an increasingly important role.

Internet technologies cross into the BAS field in a few ways. First, the Internet as a transportation medium is employed for remote access or integration of BASs located long distances apart, basically to replace the use of a telephone network based on modem technology for the same application. The second application situation is similar to the intranet concept, with Internet technologies and protocols, such as TCP/IP, used in building up the

BAS network itself no matter whether it is connected to the Internet or not. Nowadays, the use of Internet technologies in BAS is extended to the upper layers of the OSI reference model. The higher-layer protocols and technologies, such as HTTP and Internet browsers, are used for network integration (including the integration of BAS networks themselves and integration of distant BASs over the Internet) and management software development.

6.5 Use of Internet technologies at automation level

Internet protocols and technologies are used in BAS to varying degrees. The lowest degree of integration application at automation level is the case when the protocol itself is not compatible with Internet Protocol and the Internet is employed as the transmission medium only. Where protocol standards are concerned, BACnet Annex H.3 is a standard specifying such a method. This method is a substitute for the old telephone service which a traditional BAS adopted as a remote access method. Another higher degree of integration application at automation level is the case where the protocol within BAS is revised to be compatible with IP, so that devices compatible with the BAS protocol can be connected directly to IP networks. BACnet Annex J is a standard specifying such a method. In this case, the BAS networks are in fact Internet/intranet.

Integrating BAS with the Internet or an extension of Internet protocols into BAS networks are trends in the development of BAS communication protocols, such as BACnet and LonWorks. As an ISO standard for BAS, the latest BACnet has good integration capability with the Internet. The next section discusses how BACnet allows the integration of BAS protocols with the Internet.

6.5.1 Internet technologies employed to connect BACnet to an IP network

For BACnet to utilize the Internet for BAS network communication, it must speak the language of the Internet known as Internet Protocol. IP by itself is little more than an envelope with 'from' and 'to' addresses and a place for a message within. For equipment to communicate on the Internet, a Transport Layer protocol must also be used. Currently, there are two primary Transport Layer protocols, Transmission Control Protocol and User Datagram Protocol (introduced in Section 6.2.2). TCP is a reliable connection-oriented transport service that provides end-to-end reliability, resequencing and flow control. The TCP/IP combination works much like a telephone call: a connection is requested, established, and then bidirectional communication follows. UDP is a connectionless 'datagram' transport service. It is used by applications that do not require the level of service of TCP, or that wish to use communication services not available from TCP, such as multicast and broadcast delivery. Since the BACnet protocol itself provides for the guaranteed delivery of packets, resequencing and flow control, it does not require the use of TCP. Therefore UDP is utilized by BACnet.

UDP/IP was added to the BACnet specification first in Annex H.3 and later with Annex J and requires specific devices or services to be available on the BACnet network. These two annexes provide two technical approaches for using Internet technologies in the first two application cases described in Section 6.4.

6.5.2 BACnet Annex H.3

Annex H.3 specifies a BACnet/IP Packet-Assembler-Disassembler (PAD) router to be placed on every BACnet network that is to be connected over an IP network to another BACnet network. The PAD does not need to be a physically distinct device and its services can be part of a device that performs other operations, such as a building controller. The other BACnet devices in the BAS network are not IP devices.

The PAD basically acts like a BACnet router with a few specific features. When receiving a BACnet message for a device on another BACnet network, a network which can be reached only through an IP inter-network, PAD puts the BACnet message into a UDP/IP message with the IP address of the PAD device on the destination BACnet network and sends the message over the IP inter-network. The receiving PAD retrieves the BACnet message and transmits the message on the local BACnet network. Since the PAD sends the message directly to a known IP of another PAD, it does not broadcast global messages the way BACnet routers do. In fact, the use of broadcasting IP messages is not usually allowed on IP inter-networks due to the increased traffic and processing. This does require that the PAD maintain a table of its peer PAD devices.

The BACnet devices originating and receiving the messages are unaware

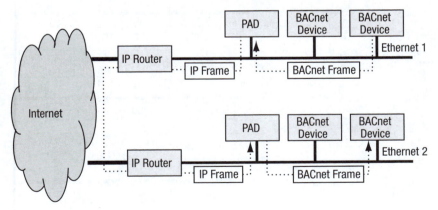

Figure 6.8 Two BACnet networks connected via the Internet using Annex H.3 PAD devices.

of the IP inter-network and communicate with the PAD devices as if they were ordinary BACnet routers connecting BACnet networks. Figure 6.8 shows how a PAD device connects two BACnet networks together with an IP inter-network (or the Internet) in between. The PAD devices appear to the IP inter-network as a device communicating using IP, not as an IP router.

6.5.3 BACnet Annex J

Using Annex H.3 PAD devices is the simplest way to connect existing BACnet networks to an IP inter-network. One problem they have, however, is that since a table of all peer PADs should be set in each PAD, there is a certain amount of manual programming (maintaining the tables) that needs to be done to all PADs each time the configuration changes by adding or removing networks. Another problem is that the addition of a single BACnet device to a BACnet system via an IP inter-network would require either that the device itself provides the PAD service or the addition of another device that provides the PAD service. It would significantly increase the cost to add a single or small group of building controllers to a BACnet network via an IP inter-network.

With the above issues in mind, as well as others, in January 1999 the IP Working Group of ASHRAE's BACnet Standing Committee (SSPC 135) developed a more extensive protocol called BACnet/IP, which was added to the BACnet standard in Annex J of the specification. BACnet/IP has several advantages over Annex H.3:

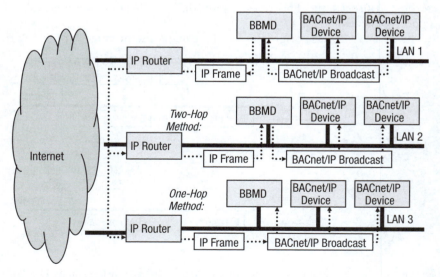

Figure 6.9 Multiple Annex J networks connected via the Internet.

- BACnet/IP is better equipped to handle BACnet broadcasts over IP than PAD devices.
- It allows devices to enter the system from anywhere in the IP inter-network.
- It supports 'native IP' BACnet devices such as unitary controllers which send and receive BACnet messages within IP frames, effectively using IP inter-networks as BACnet LANs.

Basically, BACnet/IP devices communicate using IP messages in lieu of BACnet messages, allowing the devices to be easily added to any IP inter-network. Broadcasting messages continued to be a problem and have been solved using one of the two methods outlined in Annex J called 'Multicasting' and 'BACnet/IP Broadcast Management Device' or BBMD.

In Figure 6.9, every BACnet device is an IP device, so they can be connected directly on intranet or Internet. Every BACnet subnet is an IP subnet.

6.6 Use of Internet technologies at management level

Currently, it is a trend in the BAS and IB industry to use Internet technologies at management level to develop BAS and IB systems, particularly management software and integration of BAS via the Internet. Many efficient and convenient tools, methods and packages for applications on inter-networks are available. Market pressure has seen a very quick process of selection, and the survivors are used and accepted by many (if not most) users. For instance, many BAS/IB management software packages have now been developed using HTTP and Internet browsers. This means the management functions of many other devices can be added easily into the management software packages, such as digital CCTV monitoring and control. The use of Internet technologies for developing BAS/IB management software has the following advantages:

- It allows the management software to be developed much more quickly at a low cost, as many standard functions and tools can be used and adopted directly.
- It means that management software packages developed are more open in terms of protocol and technologies. Thus it is easier for ordinary users to use them as they are similar to other software packages used in daily work or life. That is appealing to users.
- It means software packages can easily adopt the management functions of new 'third-party' devices and systems, as many of the current hardware devices are developed to be Internet compatible.

When integration of BAS with the Internet at management level is considered, current applications can be illustrated by two typical categories of integra-

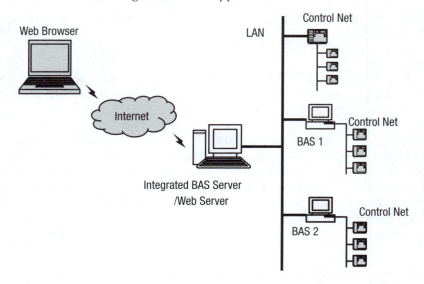

Figure 6.10 Example of BAS integration on the Internet – LAN-integrated Internet-accessible.

tion applications on the Internet, which are different in terms of the degree of integration.

The first category is the LAN-Integrated Internet-Accessible application, as illustrated in Figure 6.10. In this case, the management software communicates with various BAS software (BAS 1, BAS 2) and control net devices. All the BASs and devices are monitored and controlled by the management software. Being integrated with the management software, which acts as an 'agent', one BAS can interoperate with another BAS device. In this case, all the BASs and control devices are integrated on a LAN, very probably using Internet technologies. The management software is a web server, which means a remote user can monitor and control the BAS data points by browsing the web pages over the Internet.

This category of integration and management software can achieve integration and interoperation of BASs in the LAN. Internet access is enabled too. Remote users can monitor and control the BAS over the Internet. However, the BASs are actually integrated on LAN instead of the Internet, even although an HMI with centralized web pages may be provided for user access on the Internet.

The second category is the real 'information and services integration' on the Internet as illustrated in Figure 6.11. In this case, different BASs communicate over the Internet using middleware technology, with BASs and devices exchanging data and information. The middleware components communicate by standard protocol, such as Extensible Markup Language (XML) protocol, providing a series of Web Services to be accessible by customer's

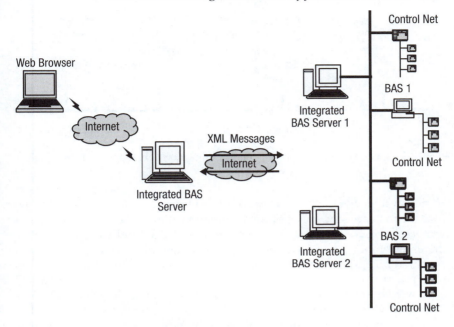

Figure 6.11 Example of BAS integration on the Internet – information and services integration using middleware technology.

applications. The applications can request data and control data points of various BASs by the Web Services method. Therefore, it is a kind of data and services integration on the Internet.

BAS software may also invoke other Web Services on the Internet to achieve specific functions. For example, a weather bureau could offer a Web Service that allows a building management system to automatically retrieve temperature forecast data for use by various control algorithms. Similarly, the building automation system itself could offer a Web Service that allows a tenant's accounting system to obtain up-to-the-minute figures on energy consumption or environmental control.

6.7 Convergence networks and total integration

Traditionally, data, video and voice control messages have been transmitted by completely different networks. Usually, the data communication network is the Internet or an intranet based on IP technology. Building automation and control message transmission is based on control networks. The voice communication network is the Public Switched Telephone Network (PSTN). In a building, specific cables are normally needed for the door phone system and public address system. Video transmission is based on the coaxial cable networks specially installed, such as for a CCTV system.

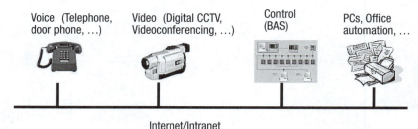

Figure 6.12 Convergence networks and total integration based on IP technologies.

However, due to the popular use of IP and Internet technologies and therefore the increasing capacity of data communication networks, all networks are beginning to converge. Total integration of BAS and IB systems was a dream not more than two decades ago, something that would provide great potential functions, services and capabilities. It is now the reality and common practice.

New technologies now make it not just possible but convenient to route voice and video phone calls over an enterprise's data network, as well as over the Internet. Voice can be transferred by VoIP (Voice on IP) technology on the Internet or intranet. As a kind of control network, BAS networks are adopting the IP network and thus can be merged conveniently into Internet/ intranet, usually an enterprise network. The video systems (CCTV) and audio system (door phone system and public address system) can also be integrated into the IP networks of BAS and therefore merged easily into the enterprise network. The enterprise network usually performs as the data network for office automation systems, and provides e-mail and Internet access, telephone network (VoIP), videoconferencing network, video on demand network and digital TV network on the same network. In such a way the total integration of building management systems can be carried out and the total integration of intelligent building systems (i.e. building automation, communication automation and office automation) is achieved easily on the basis of IP technologies and an enterprise network. Figure 6.12 provides an illustration on the convergence networks and the total integration of IB systems.

References

ANSI/ASHRAE Standard. (2007) *BACnet® – a data communication protocol for building automation and control networks*, Atlanta, Georgia: American Society of Heating, Refrigerating, and Air-Conditioning Engineers.

CISCO. (2008) *Internetworking Technology Handbook*. Available at: www.cisco. com/en/US/docs/ internetworking/technology/handbook/ito_doc.html

Cyberspace Center of HUST. (2008) *Introduction to Intranet Technologies and Products*. Available at: www.cyber.ust.hk/handbook3/hb3main.html

7 Process control, PID and adaptive control

Installation management and control are the basic tasks of building automation systems. Process control is one of the major installation management and control functions. In building systems, closed-loop control is implemented in most controlled processes. To achieve stable control of closed loops, hundreds of control algorithms have been invented. However, it has been observed that PID algorithms are used in over 80 per cent of closed control loops in industrial applications. In buildings, this percentage may be even higher.

This chapter presents the characteristics of closed-loop control, proportional control, integral control, derivative control, PID control, the tuning methods of PID loops, and digital PID control functions, as well as advanced topics related to closed-loop control including auto-tuning, gain-scheduling control and self-tuning control.

7.1 Closed control loops

Depending on what information is used to generate the control actions in a control system, all the control systems can be classified into two fundamental types, open loop and closed loop (feedback).

Figure 7.1 shows the functional block diagram of an open-loop control system. In the open-loop systems, the process output variable (i.e. controlled variable) is determined by the combined effects of the disturbance inputs and the manipulated input (i.e. energy and/or materials) which is varied by the actuation device in response to signals from the controller. The controller, according to the desired value of the controlled variable, generates a control

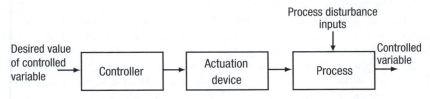

Figure 7.1 Block diagram of an open control loop.

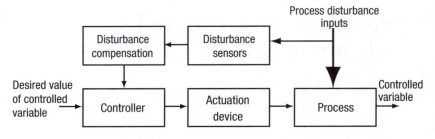

Figure 7.2 Disturbance compensated open-loop system.

signal for the actuation devices by implementing the control law or control algorithm, which is practically based on the predicted correlation between the process input and output.

Open-loop systems, without considering the effects of disturbances, as shown in Figure 7.1, can be satisfactory or acceptable if the disturbances are not great and the changes in desired values are not too severe. However, in many control systems in practice, the effects of disturbances are great or the errors in the controlled variables caused by such disturbances are serious. To reduce errors in the controlled variable in the open-loop system, the effects of the disturbances should be considered by the control scheme in generating the control signals. Implementation of such a scheme requires, as shown in Figure 7.2, that:

• the disturbances can be measured;
• the effects of the disturbances on the controlled valuable can be estimated, so that they can be compensated for.

It can be very costly and sometimes impractical to quantify the disturbances and to estimate the effects accurately. We can use the room temperature control in a heated room as an example. The heating load of the room is strongly affected by the outdoor air temperature, solar radiation, infiltration, internal heat from people, lighting and equipment, and so on. To control the output of the heat and therefore control the room temperature accurately, first, we need to measure the changes in the external and internal disturbance factors. Second, we should have an accurate mathematical model correlate these external and internal parameters on the heating load. When accurate control of the room air temperature is needed, the control system needed will be very complex, if not impossible, even for such a simple application. Certainly, open-loop systems are stable and have no control stability problems as occur in closed-loop systems.

To solve the problem of complexity, a simple solution can be very effective. That is to use the controlled variable, the final target of the system, as the reference in making the control decision. Let us consider how an operator can control the room temperature in this case.

A very simple and efficient way to regulate the room temperature is to check the actual room temperature when controlling the heater. That leads to the closed-loop or feedback control which we actually use often, even when not aware of it. In order to carry out the room temperature regulation satisfactorily using feedback control, a person should meet the following requirements. He or she must be:

- told what temperature is required for the room (set-point or desired value);
- provided with some means of adjusting the temperature (control element);
- provided with some means of observing the temperature (sensing element);
- instructed what to do to move the temperature in a desired direction (control function, negative feedback).

If the system is initially in a steady state with the required temperature achieved and some disturbance then occurs, this will cause the controlled temperature to deviate from its previous condition. The operator will not know anything of this until the thermometer shows him or her that some change has occurred. This illustrates the principle that the precision of the control can only be as good as that of the sensor. When the thermometer reading has changed the operator will then compare the current reading (the controlled variable 'y') with the required temperature (the set-point 'r') and compute the difference between them, to obtain the error signal 'e':

$$e = r - y \tag{7–1}$$

Note that the signal 'e' indicates whether the controlled temperature is too high or too low, which determines the direction of the corrective action required, i.e. whether to open up the valve or close it. The size of 'e' determines the amount of corrective action necessary. When the valve is turned in the correct direction by the correct amount, the temperature will eventually return to its original value.

It can be seen that there is a closed loop of signal or information which

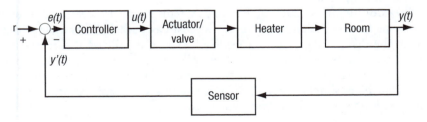

Figure 7.3 Block diagram of room temperature control using heater.

passes through the sequence: room temperature, temperature sensor, indicator, eye, brain, muscle, control valve, heating flow rate and room temperature again. The physical nature of the signal is different in various parts of the loop, but each part affects the next in turn and a change at any one point will propagate round the loop, as shown in Figure 7.3, which shows the block diagram equivalent of the operation process. It is important to notice that a feedback control loop can achieve its objective (i.e. to maintain the controlled process output at its set-point) only when the feedback of the closed loop is negative; that is, a negative feedback control loop.

A control system is defined as an interconnection of components forming a system configuration that will provide a desired system response. Because a desired system response is known, a signal, somehow, proportional to the error between the desired and the actual response, is generated. The utilization of this signal to control the process results in a closed-loop sequence of operations that is called a feedback system.

In this example, we find that the introduction of the feedback control dramatically reduced the complexity of open-loop control systems and the process output can be controlled accurately at the desired value. We may summarize the benefits of using a feedback control system as follows:

- ease of control and adjustment of the transient response of the system;
- great reduction in external disturbances (except those associated with sensors) on the controlled variables;
- greater reduction of steady-state errors of the system;
- a decrease in the sensitivity of the system to variations in the parameters of the process or tolerant variations (due to wear, age and environmental effects).

The addition of feedback to a control system results in the advantages outlined above. However, it is natural that these advantages have an attendant cost. The cost of feedback is the introduction of the possibility of instability, which is caused by the overcorrection of the process input as the results of delay and component dynamics. While the open-loop system is stable, the closed-loop system may not always be stable. The instability problem of the closed-loop control system is often an important and difficult matter in control system design.

The addition of feedback to dynamic systems results in several additional problems for the designer. However, in most cases, the advantages far outweigh the disadvantages, and a feedback system is utilized. Therefore it is necessary to consider the additional complexity and the problem of stability when designing a control. Finally, we should keep in mind that if an open-loop control, of simple design, for a process can achieve satisfactory accuracy, an open-loop system should be considered first.

7.2 Proportional control

7.2.1 The operation of proportional control

Figure 7.4 shows the block diagram of a simple control system. The block 'controller' provides a transfer function whose input signal is the error, the difference between the reference signal and the feedback signal. The controller must operate upon this input in some way so as to generate a suitable output signal to feed to the valve or other final control element. In the following, the commonly used types of control functions (PID) are discussed. *PID (proportional, integral and derivative)* controls are simple in principle. They have played a very important role in the progress of automatic control and are used as the most popular functions in process control applied in various forms.

If a person is taking the role of control output decision based on the input (error), he or she will naturally set a low output when the error (input) is small and set a high output when the error is large. Such experience can be regarded as proportional control. It is the simplest relationship between the control output ($u(t)$) and input ($e(t)$), whereby the changes in input are multiplied by a suitable constant value (gain, K_p) to produce the output change:

$$u(t) = K_p{}^*e(t)+ u_o \qquad (7\text{--}2)$$

where, u_o is a nominal control output without error. The gain constant cannot normally be a pure number since the nature of the output signal is usually different from that of the input. The actual unit of the proportional gain varies depending on the units of the controlled variable and control output.

Generally, the controller gain K_p is an adjustable parameter whose value can be chosen and set by the operator. In theory the output signal may apparently take any value but in practice its range is limited by the finite range of controller output or finite range of the final control element: a valve or damper can go only from fully closed to fully open. This fact can be regarded as a saturation effect. It produces the practical input/output curves as shown in Figure 7.5. The curves give the appropriate action for a heating duty; for a cooling duty, normally a negative gain should be applied.

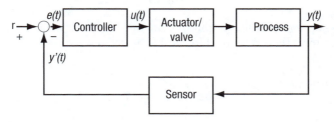

Figure 7.4 Block diagram of a simple feedback control system.

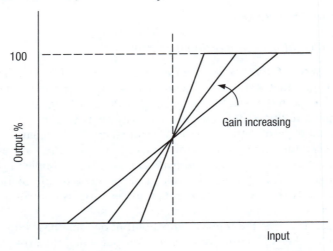

Figure 7.5 Input/output characteristics for a proportional controller with saturation by limited controller output or travel of valve.

In general, a proportional controller has to have two variables which can be set by the operator by means of front panel switches or knobs: the set-point, which is the value of the controlled variable which the controller is set to maintain, and proportional gain, which is in effect the slope of the input/output curve. A third parameter named nominal control output is often applied in the proportional controller, which can reduce the steady-state offset as discussed later in this section.

Practical controllers can easily be provided with a temperature scale and setting knob for the set-point (electronic and digital controllers may provide a digital temperature display), but the proportional gain is often replaced by an alternative, known as the *proportional band*. The proportional band is illustrated in Figure 7.6. It may be defined as the change in the controlled variable required to make the final control element move through its full operating range or stroke.

$$Proportional\ Band = \frac{Full\ Output\ Range\ of\ Controller}{Proportional\ Gain}$$

Normally, we consider the output of controller to be between 0 and 100 per cent. Therefore, their relationship can be revised as:

$$Proportional\ Gain = \frac{100\%}{Proportional\ Band}$$

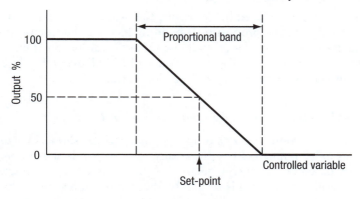

Figure 7.6 Proportional band.

Example

A proportional temperature controller controls an electric heater whose output goes from 7.0 to 4.0 kW as the controlled temperature rises from 10 to 20°C. The output of the heater can be controlled between 2 and 10 kW. Find the proportional gain and proportional band of the controller.

The proportional gain (slope of input/output characteristic) of the controller is:

$$(7.0–4.0)/(20–10) = 0.3 \ (kW/K)$$

The proportional band is given by:

$$\frac{\text{Full Output Range of Controller}}{\text{Proportional Band}} = \frac{(10-2)\ kW}{0.3\ kW/K} = 26.67\ (K)$$

One important characteristic of proportional control is the existence of offset. *Offset* is a steady-state difference between the set-point and the actual value of the controlled variable, which is not what is desired. The offset depends on the load on the system and is an unavoidable consequence of proportional control action. It can be reduced by making the proportional band narrower, although this can produce unstable operation if the proportional band is reduced too much.

The settings (i.e. nominal control output) of proportional controllers are often calibrated so that the offset is zero when the output of the final control element is at 50 per cent of its maximum. This is the normal assumption, although in some cases the offset is calibrated to be zero when the output of the final control element is at its maximum value. In practice, the operator may calibrate the offset to be zero when the load is at 'nominal level' to keep the offset at a minimum.

The offset can be reduced by increasing the proportional gain of the controller (or reducing the proportional band), which will be discussed below. Unfortunately, it will be found that the range of the proportional gain is limited in any control loop. Too high a value of K_p will make the closed loop unstable as shown in the example in Section 7.2.2.

7.2.2 Steady-state response of proportional control

The heated room controlled by a proportional controller, shown in Figure 7.7, is taken as an example to study the steady-state response of a proportional controller. Based on the energy balance of the room and the characteristics of the controller, it is possible to predict the steady temperature which will exist for a given load, depending on the set-point, proportional band and nominal control output.

From the example below, it can be seen that the offset of a closed loop controlled by proportional control can be reduced if the proper value of nominal control output is selected. If the proportional gain increases, the offset can be reduced. However, the offset cannot be eliminated in practice due to the changes in working conditions, disturbance and fixed nominal control output.

Example

The wall of an office has an overall heat transfer coefficient of 0.5°kW/K. A thermostat is used to control the electrical heater installed, which performs the P-control determined by $u(t)=u_o+K_p*e(t)$. The output of the heater is linear to the control input with maximum capacity of 12 kW.

The set-point of the indoor temperature is 24°C. If the daily average outdoor temperature in winter is 8°C, select the most suitable value of u_o in order to minimize the offset of controlled indoor temperature.

The proportional gain of the thermostat is set to be 0.4 l/K and the value of u_o selected above is used. Derive the actual room temperature when the outdoor temperature is 4°C.

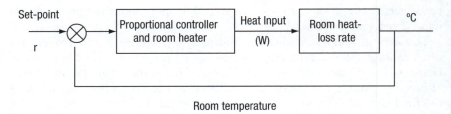

Figure 7.7 Proportional control of room temperature.

Solution

The best value of u_o should produce a 'zero' offset at nominal load (at 8°C outdoor temperature). The steady-state thermal balance of the room in this case:

$$(u_o+K_p{}^*e)^*P_{max} + UA^*(T_{atm}-T_r) = 0$$

The P control output is u_o, when the offset is zero:

$$u_o{}^*12 + 0.5^*(8-24) = 0$$

The best value of u_o is:

$$0.667 \text{ (or 66.7\%)}$$

When the room temperature is T_r, the output of P control is:

$$u = u_o+Kp^*(T_{set}-T_r)$$

The steady-state thermal balance of the room (at 4°C outdoor temperature):

$$[u_o+K_p{}^*(T_{set}-T_r)]^*P_{max} + UA^*(T_{atm}-T_r) = 0$$

$$[0.667+0.4^*(24-T_r)]^*12 + 0.5^*(4-T_r) = 0$$

The actual room temperature is the balanced temperature: *23.62°C*

7.3 Integral control

If a proportional controller can use a large gain and preserve good relative stability, system performances, including those on steady-state error, may meet requirements. However, if difficult process dynamics such as significant dead times prevent use of large gains (which is often the case for practical processes), steady-state error performance may be unacceptable. When human process operators notice the existence of steady-state errors due to changes in desired value and/or disturbance they can correct for these by changing the desired value (set-point) or the controller output until the error disappears. This is called manual reset. Integral control is a means of removing steady-state errors without the need for manual reset.

We shall see that integral control can be used by itself or in combination with other control modes. Proportional plus integral control is the most common mode. The reason proportional control suffers from steady-state errors is that a change in desired value or disturbance will require a new value of a manipulated variable in order to achieve equilibrium at the new operating conditions.

In a proportional controller, manipulated variable is proportional to *process error*, a new value of manipulated variable is possible only if non-zero error exists. We need a control that can provide any needed steady output (within its design range, of course) when its input (the process error) is zero. Although integral control is very useful for removing or reducing steady-state errors, it has the undesirable side-effects of reducing response speed and degrading stability. The reduction in speed is most readily seen in the time domain, where a step input (a sudden change) to an integrator causes a ramp output, a much more gradual change.

The integral control action (I-action) can be represented by the area under process error described by the time domain equation (7–3). From this equation, one can find that a non-zero control output, $u(t)$, can be generated by the proportional controller when the current error term, $e(\tau)$, itself is zero, providing its integral term, $\int_0^t e(\tau)d\tau$, holds a proper non-zero value.

If the current value of the integral action cannot provide the right control output to control the process output at its set-point exactly (i.e. zero error), the error term will be non-zero. This non-zero error will push the value of integral action towards the right value given sufficient time, if the working condition and disturbances remain unchanged. In practical processes, the working condition and disturbances will not remain unchanged, and correction of the control output (value of integral action for integral control) is a regular task. That results in the need for non-zero error (for the correction) and imperfect control. However, for a properly tuned control loop, the error needed for the necessary correction will not be large most of the time and the process output will be therefore controlled near its set-point as well.

$$u(t) = K_i \int_0^t e(\tau)d\tau \qquad\qquad (7-3)$$

Figure 7.8 shows the correction mechanism of integral control responding to the needs in changing (increasing and reducing) the control action. At the starting period (Period I), the error is usually large and the I-action increases quickly. When the error becomes smaller, the increasing speed of I-action becomes lower and the closed-loop system is approaching its steady state. The I-action stops changing when the error reaches zero (at Point A) and remains unchanged (Period II) until a change is needed. At Point B, there is a need to increase the control action; the error deviates from zero (positive error) due to the current control action not being sufficient. The I-action is tuned to increase as the result of increasing accumulation of integral items due to the existence of the positive error in Period III until the error reaches zero (Point C). At Point D, there is a need to reduce the control action; the error deviates from zero again (negative error) due to the current control action being too large. The I-action is tuned to reduce as the result of reducing accumulation of integral items due to the existence of the negative error in Period V until the error reaches zero (Point E). In a properly tuned control

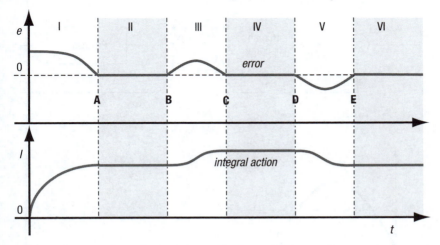

Figure 7.8 Correction mechanism of integral control responding to the needs in changing the control action.

loop, such correction is an ongoing practice resulting in the process variable being controlled within a small range around its set-point.

Since integral control is usually used together with proportional control, the actual integral action appears in the format as below:

$$u(t) = K_p \left[e(t) + \frac{1}{T_i} \int_0^t e(\tau)d\tau \right] \tag{7-4}$$

The integral action becomes:

$$I(t) = \frac{K_p}{T_i} \int_0^t e(\tau)d\tau \tag{7-5}$$

where, T_i is called the *integral time* of integral control. Integral time is actually a coefficient which has the unit of time. It determines how strong the integral control action is. A smaller value of integral time gives a stronger degree of integral control. In practice, integral control is often used together with proportional control.

7.4 Derivative control

Although the proportional and integral controls described in previous sections can each be used as the sole function in a practical controller, we will see that the various derivative controls are always used in combination with some more basic control law. This is because a derivative function produces no corrective effect for any constant error, no matter how large, and would

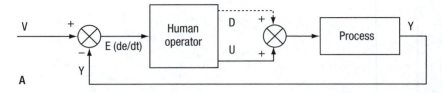

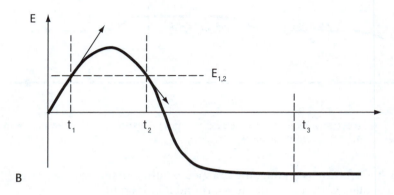

Figure 7.9 Possible genesis of derivative control.

therefore allow uncontrolled steady-state errors. Thus the discussion in this section cannot consider derivative functions in isolation. They will always be considered as supplementing some other function.

Although derivative functions are helpful in solving a variety of control design problems, one of their most important contributions is in system stability improvement. If absolute or relative stability is the problem, a suitable derivative control mode is often the answer.

This stabilization or 'damping' aspect of derivative control can be easily understood qualitatively from the following discussion. Just as the invention of integral control may have been stimulated by human process operators' desire to automate their task of manual reset, derivative control may first have been devised as a simulation of human response to changing error signals. In Figure 7.9, we assume a human process operator is given a graphical display of system error E and has the task of changing manipulated variable U so as to keep E close to zero.

Let us consider '*If I were the operator, would I produce the same value of U at time t_1 as at time t_2?*' (Note that a proportional controller would do exactly that.) Most people agree that a higher corrective effort seems appropriate at t_1 and a lesser one at t_2, since at t_1 the error is $E_{1,2}$ and increasing, whereas at t_2 it is also $E_{1,2}$ but decreasing. That is, the human eye and brain senses not only observe the current value of the graph's curve but also its trend or slope. Slope is clearly dE/dt, so to mechanize this desirable human response we need a controller sensitive to error derivative. Such a control can, however, not be used alone since it is not sensitive to steady-state errors of any size, as at t_3, thus a combination of proportional plus derivative, for example, makes sense.

The derivative control action can be represented by a function of the error slope as the time domain equation:

$$D(t) = K_D * de(t)/dt \qquad\qquad (7\text{--}6)$$

Since derivative control is usually used together with proportional control, the actual derivative action appears in the format as below:

$$u(t) = K_p[e(t)+T_d * de(t)/dt] \qquad\qquad (7\text{--}7)$$

The derivative action then becomes:

$$D(t) = (K_p * T_d) * de(t)/dt \qquad\qquad (7\text{--}8)$$

where T_d is called *derivative time* of derivative control. Derivative time is actually a coefficient which has the unit of time. It determines how strong the derivative control action is. A larger value of derivative time gives a stronger degree of derivative control. In practice, derivative control is often used together with proportional control or proportional control plus integral control.

In addition to the stability argument already discussed, derivative functions may also offer improvement in speed of response. P+I control can provide satisfactory control stability for most process controls in buildings. A D controller should be used very carefully since an incorrect parameter may have a negative effect on the control loop stability. Different manufacturers implement very different D control algorithms in their controller, which also makes it difficult for operators to tune the D control parameters.

7.5 Proportional, integral and derivative functions

It appears that, when any given value of set-point is set on a proportional controller, the controlled variable can be controlled at that set-point value only at one particular load. In all other cases there is a non-zero offset and a difference exists between the set-point and the controlled variable. In many cases the proportional band can be set relatively narrow and the offset is acceptably small. In cases where the load is steady for long periods it may be possible to adjust the set-point so that the controlled value becomes equal to the desired value, but this is not often convenient.

In a closed control loop, the proportional band can be reduced to reduce the offset. But this is equivalent to increasing the gain as shown in the previous sections. This can produce oscillation and instability. Such effects set limits on the accuracy of control with proportional action only. If better performance is required, it is necessary to resort to some alternative or additional means of control action.

Control functions are available to provide additions to the basic proportional control in order to remove the offset and speed up the response. They are integral control and derivative control. When integral control is added to proportional control, the offset can be eliminated. The integral function generates a control signal which is proportional to the time integral of the error signal. A continuous offset will therefore eventually produce a corrective action, which will continue to accumulate until the error has reached zero.

In order to speed up the response of the system to changes in load, a derivative function can also be added to the proportional controller. This generates a control signal which is proportional to the time derivative or instantaneous rate of change of the error. This has no effect on the offset, but it allows the use of larger proportional gain and produces earlier corrective action, taking the trend of changes into account.

Most digital controllers combine all three types of action (proportional, integral and derivative, P+I+D) to incorporate their advantages. The relative strengths of the control actions have to be adjusted to match the characteristics and response of the process which is being controlled. The proportional gain constant K_p determines the sensitivity of the proportional action but since the integral and derivative actions are time-related, they are adjusted by reference to so-called action times: the integral (action) time T_i and derivative (action) time T_d.

Proportional plus integral plus derivative (P+I+D) function can be represented by the time domain equation:

$$u(t) = K_p \left[e(t) + \frac{1}{T_i} \int_0^t e(\tau)d\tau + T_d \frac{de(t)}{dt} \right] \tag{7–9}$$

This is the most classical form of PID algorithm called *ideal non-interacting PID controller* or *ISA algorithm*. The transfer function of P+I+D control is:

$$PID(s) = K_p \left(1 + 1/T_i s + T_d s \right) \tag{7–10}$$

The examples presented below are also commonly used. They are named *ideal parallel PID controller* (7–11) and *interacting PID controller* (7–12). In practice, attention should be paid to the form of PID algorithms used for tuning a practical controller since PID controllers of different forms are very different in selecting parameters.

$$u(t) = K_p e(t) + \frac{1}{K_i} \int_0^t e(\tau)d\tau + K_d \frac{de(\tau)}{dt} \tag{7–11}$$

$$u(t) = K_p \left[e(t) + \frac{1}{T_i} \int_0^t e(\tau)d\tau \right]\left[1 + T_d \frac{de(\tau)}{d\tau} \right] \qquad (7\text{--}12)$$

where, K_p is gain, T_i and K_i are integral settings, and T_p and K_d are derivative settings of the controller.

In summary, integral function tends to destabilize the system; derivative control function tends to reinforce the stability. Although integral function is often introduced in order to reduce or remove the offset of proportional control and to give more accurate control, its effect may make the system oscillate and worsen the control. Although in theory the addition of derivative function to a proportional controller will not remove the offset, it will nevertheless allow a narrower proportional band to be chosen so that offset may be reduced to an acceptable value. It is extremely important to choose suitable settings for K_p, T_i and T_d. The next section describes some simple empirical methods with which an approximate initial choice can be made.

7.6 Tuning of PID control loops

In order to give some guidance on suitable initial settings for controllers, which will give reasonably good control without being either too sluggish or too oscillatory, a number of simple tests have been developed. These were more specifically aimed at the process industries, but there is no reason why they should not be used for other applications such as building services.

The assumption is that there is a closed-loop negative feedback control process as in Figure 7.10 where the characteristics of the actuator/valve, process and sensor are fixed and it is necessary to set the controller so as to match these characteristics satisfactorily.

Two basic methods are discussed below: the open-loop test method (or process reaction curve method) and closed-loop test method (or ultimate cycling method). The open-loop test method may need to disconnect the connection of the control loop, while the modern digital controller usually allows us to 'disconnect' the closed control loop conveniently by software setting or configuration without disconnecting the hardware connection. The closed-loop test may introduce strong oscillation which may be harmful to the installation. The choice of the test method depends on the ease of use and possibility of implementation in practical cases.

7.6.1 Open-loop test method

In this method the controller is isolated and the closed-control loop is disconnected. The test is carried out on the remainder of the control loop. It must be arranged so that a signal can be injected in place of the controller output and another signal, which would enter the controller input, can be observed or recorded (see Figure 7.10). A sustained step input of magnitude

X is applied to the valve, and the response at Y_2 is observed and recorded, as in Figure 7.11.

In most cases, the response will follow the general form of the S-shaped curve shown in Figure 7.11, which is the resultant response of all the lags, delays, time constants and gain factors in the controlled system. From the experimental curve three values are measured, corresponding to a pure delay L, a single time constant T and a steady-state gain factor Y/X. The equivalent overall transfer function of the open loop is therefore:

$$\frac{(Y \, / \, X) \, exp \, (-Ls)}{1 + Ts}$$

It is then possible to calculate the necessary controller settings to give a reasonably good control performance. There are many proposals relating to the suggested PID settings with the above system characteristic parameters obtained from the test. Two well-known sets of such proposals are listed in Tables 7.1 and 7.2.

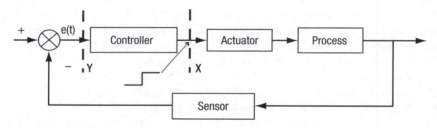

Figure 7.10 System for empirical choice of controller settings.

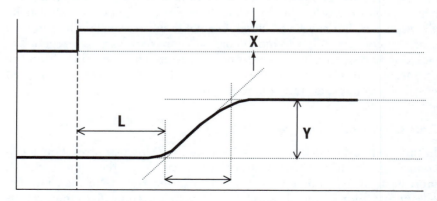

Figure 7.11 Input and output waveforms for open-loop test method.

Table 7.1 Ziegler and Nichols settings (*A=X/Y, B=T/L*; see Figure 7.11)

Type of Controller	K_p	T_i	T_d
P	A*B	–	–
P+I	0.9*A*B	3.3*L	–
P+I+D	1.2*A*B	2*L	0.5*L

Table 7.2 Cohen and Coon settings (*A, B* as before, *R=L/T*; see Figure 7.11)

Type of controller	K_p	T_i	T_d
P	AB(1+R/3)	–	–
P+I	AB(1.1+R/12)	$\frac{L(30+3R)}{(9+20R)}$	–
P+D	AB(1.25+R/6)	–	$\frac{L(6-2R)}{(22+3R)}$
P+I+D	AB(1.33+R/4)	$\frac{L(32+6R)}{(13+8R)}$	4L/(11+2R)

7.6.2 Closed-loop test method

An alternative empirical test which has been suggested is to allow the controller to remain connected in the control loop but to decrease the proportional band progressively (corresponding to increasing the gain K_p) until the system just sustains steady oscillations of constant amplitude as shown in Figure 7.12. The test procedure is as follows. First, the integral and derivative functions must be disabled in the controller or their effects should be reduced to a minimum by setting T_i to its maximum value and T_d to its minimum value. Then the proportional gain is increased progressively while observing the response of the control loop. The smallest value of K_p

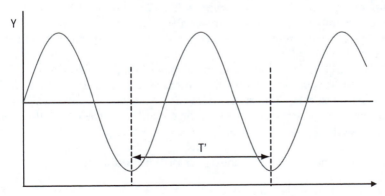

Figure 7.12 Process response in closed-loop test method.

Table 7.3 Controller settings for the closed-loop test method

Type of controller	K_p	T_i	T_d
P	$0.5K_p$'	–	–
P+I	$0.45K_p$'	$0.8T$'	–
P+I+D	$0.6K_p$'	$0.5T$'	$0.125T$'

which just produces a steady oscillation without dying away is designated K_p'. The period of the oscillations is measured (see Figure 7.12) and is then designated T'. The recommended controller settings are then related to K_p' and T' and are set out in Table 7.3.

7.7 Digital PID and direct digital control (DDC)

The basic function of a digital controller (outstation) is to perform direct digital control. *Direct digital control* is often used in place of conventional pneumatic or electric local control loops. There are several industry accepted definitions of DDC. It can be defined as '*a control loop in which a digital controller periodically updates a process as a function of a set of measured control variables and a given set of control algorithms*'. In fact, DDC is the local control loop in which the digital controller is used.

As in conventional types of controllers used in buildings, PID also plays a very important role in most control loops controlled by a digital controller. As in conventional types of controllers, there are no industry standards for PID controllers. The PID algorithms used in actual controllers from different manufacturers differ greatly. Manufacturers also change both names and units. Therefore, one must be very careful regarding the actual algorithms used and the definitions of the parameters to be tuned when using a controller obtained from the market. Even if two digital controllers from different manufacturers have PID algorithms which can be described by the same analogue form of PID function, the control performances might be notably different due to different discrete form representations of those PID functions and sampling periods.

However, PID algorithms from most manufacturers fit into one of three major classifications: interacting, non-interacting and parallel. Manufacturers also differ as to their names for these categories. So the only way to really tell which one you have is to look at the equation for the controller. In ideal form, these three categories are:

- Ideal non-interacting PID controller or ISA algorithm:

$$u(t) = K_p \left[e(t) + \frac{1}{T_i} \int e(t)dt + T_d \frac{de}{dt} \right] \qquad (7\text{--}13)$$

- Ideal parallel PID controller:

$$u(t) = K_p e(t) + \frac{1}{I_p} \int e(t)dt + D_p \frac{de}{dt} \tag{7-14}$$

- Interacting PID controller:

$$u(t) = K_p \left[e(t) + \frac{1}{T_i} \int e(t)dt + T_d \frac{de}{dt} \right] \left[1 + T_d \frac{de}{dt} \right] \tag{7-15}$$

where K_p is gain, T_i and I_p are integral settings of the controller, and T_d and D_p are derivative settings of the controller.

Figure 7.13 shows the block diagram of a typical DDC loop. A digital controller normally has a number of input channels and output channels. Therefore, usually more than one DDC loop is controlled by one digital controller.

Figure 7.14 shows the signals passing through a DDC loop. It can be assumed that all the numbers that enter or leave the computer do so at the same fixed period T, called sampling period or sampling time. The variables $r(kT)$, $m(kT)$ and $u(kT)$ are sequences of discrete signals in contrast to $m(t)$, $c(t)$ and $y(t)$, which are continuous functions of time. Data obtained for the system variables only at discrete intervals is called sampled data.

If the sampling period chosen is very small compared to the time constant of the process, the system is essentially continuous, and the methods used

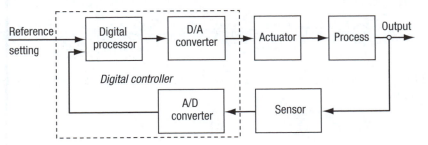

Figure 7.13 Block diagram of a typical DDC loop.

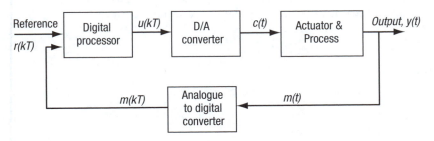

Figure 7.14 Signals passing through a DDC loop.

in the analogue control system are valid. If the sampling period is of the same magnitude as the time constants of the processes, the sampling period affects the closed-loop performance, which may occur in practice. There are theories and methods developed to analyze the stability problems of control loops of discrete signals. However, in this case one can still get useful ideas on the stability of DDC loop using the theories and methods used in analogue control systems.

A digital PID control has to perform the control function based on the discrete signals obtained from the input unit of the controller. If we have the error signals at the sampling instants as a series of discrete data:

$$e_0, e_1, e_2, e_3, \dots \dots, e_{k-2}, e_{k-1}, e_k$$

the PID actions of the ISA algorithm as in Equation (7–13) in the time (sampling) instant K can be computed as below.

P-action:

$$K_p e(t) \Rightarrow K_p e_k \tag{7–16}$$

I-action:

$$\frac{K_p}{T_i} \int e(t)dt \Rightarrow \frac{K_p T}{2T_i} \sum_{j-1}^{k} (e_{j-1} + e_j) \tag{7–17}$$

D-action:

$$K_p T_d \frac{de(t)}{dt} \Rightarrow K_p T_d \frac{e_k - e_{k-1}}{T_{samp}} \left[\text{or: } K_p T_d \frac{e_k - 2e_{k-1} + e_{k-2}}{T_{samp}} \right] \tag{7–18}$$

The PID output in sampling case K therefore can be computed by equation:

$$U_k = K_p e_k + \frac{K_p T}{2T_i} \sum_{j-1}^{k} (e_{j-1} + e_j) + K_p T_d \frac{e_k - e_{k-1}}{T} \tag{7–19}$$

The I-action often uses incremental form such as Equation (7–20), which means only two historic items of data need to be recorded.

$$I_k = I_{k-1} + \frac{K_p T}{2T_i} (e_{k-1} + e_k) \tag{7–20}$$

The PID output is then computed by equation:

$$U_k = K_p e_k + I_{k-1} + \frac{K_p T}{2T_i} (e_{k-1} + e_k) + K_p T_d \frac{e_k - e_{k-1}}{T} \tag{7–21}$$

Table 7.4 An example of program code of PID function routine

```
FUNCTION PID(P,Tset,GAIN,Ti,Td,Iterm0,Dterm0,ERR0,TIMEB,TIME,Pte
rmN)
    Real Iterm,Iterm0
    Logical NOITERM,NODTERM
C Computing error
    ERR = Tset-P
C Sampling interval
    Tsamp = TIME-TIMEB
C Proportional action
    Pterm = GAIN*ERR+PtermN
    Pterm = MAX(-100.0,MIN(125.0,PTERM))
C Integral action
    NOITERM=(Ti.LE.0.0).OR.(Ti.GT.9E+4)
    IF (NOITERM) THEN
    Iterm = 0.0
    ELSEIF ((Pterm+Iterm0).GT.100.0) THEN
    Iterm = Iterm0
    ELSEIF ((Pterm+Iterm0).LT.0.0) THEN
    Iterm = Iterm0
    ELSE
    Iterm = Iterm0+(((ERR0+ERR)/2.0)*Tsamp/Ti)*GAIN
    ENDIF
C Differential action
    NODTERM = (Td/Tsamp).LT.1.0E-3
    IF(NODTERM) THEN
    Dterm = 0.0
    ELSE
    Dterm = GAIN*Td*(ERR-ERR0)/Tsamp
    Dterm = (Dterm+Dterm0)/2.0
    Dterm = MAX(-25.0,MIN(25.0,Dterm))
    ENDIF
C Controller output
    PID = Pterm+Iterm+Dterm
    PID = MIN(100.0,MAX(0.0,PID))
C Update previous integral and derivative actions
    Iterm0 = Iterm
    Dterm0 = Dterm
    ERR0 = ERR
    RETURN
    END
```

As an example of practical utilization of PID algorithms in a digital controller, the program code (Fortran subroutine) is listed in Table 7.4, which performs the ISA algorithms.

In this routine, the proportional action is limited within the range between -100% and $+125\%$; the derivative action uses the average of the current and previous step in order to increase the reliability and it is limited within the range -25% and $+25\%$; the PID output is finally limited to between 0 and 100%. The symbols used in the subroutine are described in Table 7.5.

Table 7.5 Description of symbols used

C	Current value of integral action (return)
Dterm	Current value of derivative action
Dterm0	Derivative action from previous sampling instant (entry); or, current value of derivative action (return)
ERR	Current value of process variable error (return)
ERR0	Process variable error from previous sampling instant (entry); or, current value of process variable error (return)
GAIN	Proportional gain (%)
Iterm	Integral action of current sampling instant
Iterm0	Integral action from previous sampling instant (entry); or, current value of integral action (return)
PID	Controller output (0–100%)
P	Process variable
Pterm	Current value of proportional action
PtermN	Nominal output of proportional action with zero error
Td	Derivative time (second)
Ti	Integral time (second)
TIMEB	Time of previous sampling instant (second)
TIME	Current time (second)
Tsamp	Sampling period
Tset	Set-point

7.8 Introduction to adaptive control

In everyday language, 'to adapt' means to change behaviour to conform to new circumstances. Conceptually, an adaptive controller is thus a controller that can modify its behaviour in response to changes in the dynamics of the process and the character of disturbances. It was suggested in an early symposium in 1961: 'An adaptive system is any physical system that has been designed with adaptive viewpoint'. An adaptive controller can be defined as *a controller with adjustable parameters and a mechanism for adjusting the parameters*.

The reason adaptive control is needed is that the constant parameters of a controller might provide satisfactory performance in one condition but not in another due to significant changes in the nature of the system. In this case, adaptive control provides a solution. Sometimes, it is used to avoid the difficulty of tuning. Typical adaptive schemes in application are as follows. The basic ideas of the first three techniques, which are used in building process control, are introduced below.

- *Auto-tuning*
- *Gain scheduling*
- *Self-tuning regulator*
- *Model-reference adaptive system*
- *Stochastic adaptive control.*

7.8.1 Auto-tuning

Some adaptive schemes require a priori information about the process dynamics. For users it would be ideal to have an auto-tuning function in which the controller can be tuned simply by pushing a button. Auto-tuning techniques are developed for this purpose. Several manufacturers were forced to introduce a pre-tune mode to help in obtaining the required prior information. Auto-tuning of PID controllers is a typical industrial application of auto-tuning. The importance of a priori information also appeared in connection with attempts to develop techniques for automatic tuning of simple PID regulators. Such regulators, which are standard control stations for industrial and building automation, are used to control systems with a wide range of time constants.

Although conventional adaptive schemes seemed to be ideal tools to provide automatic tuning, they were found to be inadequate because they required prior knowledge of time scales. Special techniques for automatic tuning of simple regulators were therefore developed. These techniques are also useful for providing pre-tuning of more complicated adaptive systems. An overview of industrial PID controllers with auto-tuning is given below.

Several ways to carry out auto-tuning have been proposed. The most common method is to conduct a simple test on the process. The test can be done in open loop or closed loop as described in Section 7.6. In the open-loop tests the input of the process is excited by a pulse or a couple of steps. A simple process model, for instance of second order, is then estimated by using recursive least squares or some other recursive estimation method. If a second-order process model is estimated, the PID controller can be used to make pole placement.

The speed and the damping of the system are then the design parameters. A popular design method is to choose the controller zeros such that they cancel the two process poles. This gives good responses to set-point changes, while the response to load disturbances is determined by the open-loop dynamics. The transient response method for automatic tuning of PID controllers is used in products in the market. It is used also for pre-tuning in adaptive controllers in some products.

The tuning experiments can also be done in closed loop. A typical example of this is the self-oscillating method of Ziegler and Nichols or its variants. The relay auto-tuner based on self-oscillation is used in some products. In these regulators tuning is initiated simply by pushing the tuning button. One advantage of making tests in a closed loop is that the output of the process may be kept within reasonable bounds, which can be difficult for processes with integrators if the experiment is done in an open loop.

The auto-tuning function is often a built-in feature in standard standalone PID controllers. Automatic tuning can also be done by using external equipment. The tuner is then connected to the process and performs a test, usually in open loop. The tuner then suggests parameter settings, which are

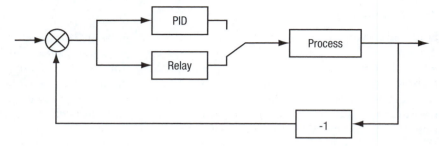

Figure 7.15 Conceptual diagram of a relay auto-tuner.

transferred to the PID controller either manually or automatically. Since the external tuner must be able to work with PID controllers from different manufacturers, it is important that the tuner have detailed information about the implementation of the PID algorithm in specific cases.

Another method for auto-tuning is to use an expert system to tune the controller. This is done during normal operation of the process. The expert system waits for set-point changes or major load disturbances and then evaluates the performance of the closed-loop system. Properties such as damping, period of oscillation and static gain are estimated. The controller parameters are then changed according to the built-in rules, which simulate the behaviour of an experienced control engineer.

Ziegler and Nichols have devised a very simple heuristic method for deter-mining the parameters of a PID controller based on the critical gain and the critical period. Systems with better damping can be obtained by using modi-fied empirical correlations obtained by various people. A modified method of this type is ideally matched to the determination of K_p and T_i, by the relay method. This gives the relay auto-tuner shown in Figure 7.15. When tun-ing is demanded, the switch is set to T, which means that relay feedback is activated and the PID regulator is disconnected. When a stable limit cycle is established, the PID parameters are computed, and the PID controller is then connected to the process. Naturally, the method will not work for some systems. First, there will not be unique limit cycle oscillations for an arbitrary transfer function. Second, PID control is not appropriate for all processes. Relay auto-tuning has empirically been found to work well for a large class of systems encountered in process control.

7.8.2 Gain scheduling

In many situations it is known how the dynamics of a process change with the operating conditions of the process. One source for the change in dynam-ics may be nonlinearities that are known. It is then possible to change the parameters of the controller by monitoring the operating conditions of the process. This idea is called *gain scheduling*, since the scheme was originally used to accommodate changes in process gain only. Gain scheduling is a special type of nonlinear feedback. It has a linear controller whose parameters are changed as a function of operating conditions in a pre-programmed way. The idea of relating the controller parameters to auxiliary variables is old, but the hardware needed to implement it easily was not available until digital controllers came into common use recently. Gain scheduling of analogue type has thus been used only in special cases, such as in autopilots for high-performance aircraft. Gain scheduling is now easy to implement in computer-controlled systems.

Gain scheduling based on measurements of operating conditions of the process is often a good way to compensate for variations in process parameters or known nonlinearities of the process. It is controversial whether a system with gain scheduling should be considered an adaptive system or not, because the parameters are changed in an open-loop or pre-programmed manner. If we use the informal definition of adaptive controllers given above, gain scheduling can be regarded as an adaptive controller. Gain scheduling is a very useful technique for reducing the effects of parameter variations. In fact, there are many building and commercial process control systems in which gain scheduling can be used to compensate for static and dynamic nonlinearities. Split-range controllers that use different sets of parameters for different ranges of the process output can be regarded as a special type of gain-scheduling controller.

In practical processes, it is sometimes possible to find auxiliary variables that correlate well with the changes in process dynamics. It is then possible to reduce the effects of parameter variations simply by changing the parameters of the controller as functions of these auxiliary variables as shown in Figure 7.16. Gain scheduling can thus be viewed as a feedback control system in which the feedback gains are adjusted by using feed forward compensation.

A key problem in the design of systems with gain scheduling is to find suitable scheduling variables. This is normally done on the basis of knowledge of the physics of a system. When scheduling variables have been determined, the controller parameters are calculated at a number of operating conditions by using some suitable design method. The controller is thus tuned or

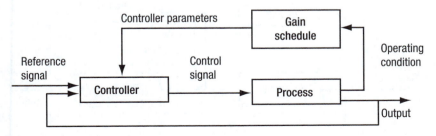

Figure 7.16 Conceptual diagram of gain-scheduling control.

calibrated for each operating condition. The stability and performance of the system are typically evaluated by simulation. Particular attention is given to the transition between different operating conditions. The number of entries in the scheduling table can be increased if necessary. However, there is no feedback from the performance of the closed-loop system to the controller parameters.

One drawback of gain scheduling is that it is an open-loop compensation. There is no feedback to compensate for an incorrect schedule. Another drawback of gain scheduling is that the design may be time-consuming. The controller parameters must be determined for many operating conditions, and the performance must be checked in tests and commissioning.

Gain scheduling has the advantage that the controller parameters can be changed very quickly in response to process changes. Since no estimation of parameters occurs, the limiting factors depend on how quickly the auxiliary measurements respond to process changes.

7.8.3 Self-tuning controller

Development of a control system involves many tasks such as modelling, design of a control law, implementation and validation. The *self-tuning regulator* (STR) or self-tuning controller attempts to automate several of these tasks. This is illustrated in Figure 7.17, which shows a block diagram of a process with a self-tuning controller. It is assumed that the structure of a process model is specified. Parameters of the model are estimated online, and the block labelled '*Estimation*' in the figure gives an estimate of the process parameters, which is the essential function block of a self-tuning controller. It is a recursive estimator. The block labelled '*Controller design*' contains computations that are required to perform a design of a controller with a specified method and a few design parameters that can be chosen externally.

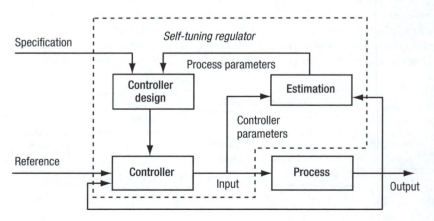

Figure 7.17 Conceptual diagram of self-tuning controller.

The design problem is called the underlying design problem for systems with known parameters. The block labelled '*Controller*' is an implementation of the controller whose parameters are obtained from the control design.

The main reason for using an adaptive controller is that the process or its environment is changing continuously. It is difficult to analyze such systems. To simplify the problem, it can be assumed that the process has constant but unknown parameters. The term self-tuning was used to express the property that the controller parameters converge to the controller that was designed if the process was known. An interesting result was that this could happen even if the model structure was incorrect.

The tasks shown in the block diagram can be performed in many different ways. There are many possible choices of model and controller structures. Estimation can be performed continuously or in batches. In digital implementations, which are most common, different sampling rates can be used for the controller and the estimator. It is also possible to use hybrid schemes in which control is performed continuously and the parameters are updated discretely. Parameter estimation can be done in many ways. There is also a large variety of techniques that can be used for control system design. It is also possible to consider nonlinear models and nonlinear design techniques. Although many estimation methods will provide estimates of parameter uncertainties, these are typically not used in the control design. The estimated parameters are treated as if they are true in designing the controller. This is called the certainty equivalence principle.

The choice of model structure and its parameterization are important issues for self-tuning controllers. A straightforward approach is to estimate the parameters of the transfer function of the process. This gives an indirect adaptive algorithm. The controller parameters are not updated directly, but rather indirectly via the estimation of the process model.

References

Åström, K. J. and Hägglund, T. (1995) *PID controllers: theory, design and tuning*, Instrument Society of America: Research Triangle Park, North Carolina.

Åström, K. J. and Wittenmark, B. (1995) *Adaptive Control*, Reading, Massachusetts: Addison-Wesley Publishing Company.

Letherman, K. M. (1981) *Automatic Controls for Heating and Air-conditioning: principles and applications*, Oxford: Pergamon Press.

8 Control and optimization of air-conditioning systems

There are several factors that influence thermal comfort: air temperature, air velocity, relative humidity, radiant environment, clothing and activity level. Air temperature is the most common measure of comfort, and the one that is most widely understood. An air-conditioning system provides the controlled environment in which the following parameters are maintained within desired ranges: temperature, humidity, air distribution and indoor air quality in order to build up a comfortable and healthy indoor environment for people to work or live in. Thermal comfort and minimum health requirements must be achieved by the basic controls of the air-conditioning system, while the optimal control of the systems aims at providing satisfied thermal comfort and indoor air quality with minimum energy input.

This chapter discusses the basic control loops, control of constant air volume (CAV) systems, control of variable air volume (VAV) systems, control of air-handling units (AHU) and VAV boxes, outdoor ventilation control and optimization, the optimal reset of supply air temperature and static pressure set-points, and the systematic optimization of air-conditioning systems. Only the air-side systems are addressed in this chapter. The control of water systems is discussed in Chapter 9. The operating principles and typical control strategies and basic instrumentations of these control systems are presented, along with the implementation of some control strategies in a simplified symbolic programming environment and a high-level-language programming environment for BAS control stations.

8.1 Typical control loops of the air-conditioning process

8.1.1 Typical air-side systems

The types of air-side air-conditioning systems include all-air systems, air-water systems, and all-water systems. Constant air-volume systems and variable air-volume systems are two major categories of *all-air systems* concerned with air flow control. Regarding ductwork design, these systems can be further divided into single-duct systems and dual-duct systems. In terms of zoning of the building, the air-side systems can be divided into single-zone systems and multiple-zone systems.

The advantages of all-air systems are as follows:

- separation of mechanical areas and major equipment from occupants;
- possibility of using outdoor air for cooling and recovery;
- flexibility to outdoor and occupancy changes;
- flexibility in design and control;
- well suited to applications requiring unusual exhaust and makeup air quantities (+/– pressurization).

The disadvantages of all-air systems are:

- they require additional duct space;
- the air balance required can be difficult;
- they require close cooperation between architectural, system and structural designers to ensure accessible terminal devices.

Fan-coil systems are the most commonly used *air-water systems*. A fan-coil system is simple in design, operation and control, and it is less expensive than all-air systems. It is flexible for zoning and requires less space for the air ductwork. However, the system is mounted in occupied spaces, which can cause noise problems. There is little flexibility in outdoor ventilation control, and free cooling is not really achieved.

It is not the objective to list all different air-side systems or to discuss their design issues in this chapter. This chapter will mainly address the operation and control of typical air systems, while the basic characteristics of the typical systems are discussed when presenting their control and operation issues.

8.1.2 Cascade control

Cascade control can be used when there are several measurement signals and one control variable. It is particularly useful when there are significant dynamics (e.g. long dead times or long time constants) between the control variable and the process variable. Tighter control can then be achieved by using an intermediate measured signal that responds faster to the control signal. Cascade control can be categorized into two kinds. One is called basic cascade control, built up by nesting the control loops as shown in the block diagram in Figure 8.1. The other is called reconfigured cascade control, which is tailored for specific applications, such as the example shown in Figure 8.2.

Basic cascade control has two loops. The inner loop is called the *secondary loop*. The outer loop is called the *primary loop*. The reason for this terminology is that the outer loop deals with the primary measured signal. It is also possible to have a cascade control with more nested loops. The performance

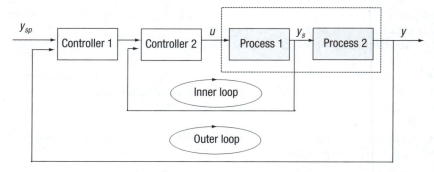

Figure 8.1 Block diagram of a system with basic cascade control.

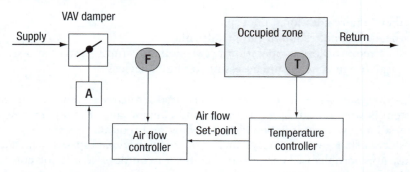

Figure 8.2 Block diagram of VAV box control with reconfigured cascade control (pressure-independent VAV box).

of a system can be improved with a number of measured signals, up to a certain limit. By dealing with the faster and slow processes (loops) in the control systems separately, the controller compromises the response speed and control stability and achieves quicker response and more stable control than that which can be achieved by using the process variable to modulate the control signal directly.

Reconfigured cascade control used for a pressure-independent VAV box is such an example; it has two loops as shown in Figure 8.2. The set-point of the direct control variable (i.e. air flow rate), which is not predefined, is calculated by a controller (of the primary loop) with measured process variable. For cooling and heating, buildings (rooms) have long time delays and long time constants, which causes significant dynamics between the control variable (flow rate) and process variable (space temperature). If space temperature is used directly for VAV damper control, seriously over-tuned or down-tuned phenomena can become obvious because of the slow thermal responses of buildings. When the space temperature is used to determine the required air flow rate with respect to the temperature set-point and the required air flow rate is compared with measured air flow rate to carry out damper PID control, the instability of the control process should be reduced greatly. Furthermore, the effects of the pressure variations on room temperature control are eliminated as the flow-control loop responds very quickly to such pressure changes before the room temperature is affected.

8.1.3 Sequential split-range control

In the previous section, cascade control is used when there is one control variable and several measured signals. Sequential split-range control (Seem 1999; Wang and Xu 2002) is used when there is one measured variable and several control variables. Systems of this type are common in air-conditioning control such as when heating and cooling are provided simultaneously. One physical device is used for heating and another for cooling (or even a third one for fresh air damper control). The heating and cooling systems often have different static and dynamic characteristics. The principle of sequential split-range control is illustrated in Figure 8.3, which shows the static relation between the measured variable and the control variables. When the temperature is very low, the heater reaches the fully open position. As the temperature increases, the heater decreases linearly until the mid-range where no heating is supplied. Similarly, cooling is applied when the temperature is above mid-range, and the cooler increases linearly to the fully open position as temperature increases further.

Sequential split-range control therefore avoids simultaneous use of heating and cooling in particular systems (zones). However, there is a critical region when switching between heating and cooling. Switching between heating and cooling control modes may cause difficulties and oscillations and consequently heating and cooling at the same time. To avoid this, a small dead zone (or band) is often adopted where neither heating nor cooling is supplied.

In sequential split-range control, there is usually one measured signal and several actuators. It is commonly used in systems for heating or cooling and ventilation. It is also useful for applications when the control variable

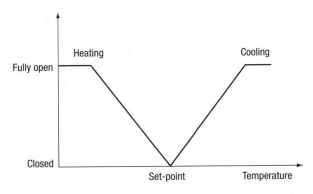

Figure 8.3 Conceptual illustration of sequential split-range control.

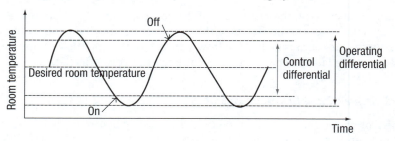

Figure 8.4 Operation of the two-position control.

operates over a very large range. The flow is then separated into parallel paths each controlled with a valve.

8.1.4 Two-position control and on/off control

Most building processes are controlled by continuous modulating control. However, there are still many processes controlled using two-position control primarily due to the low cost of the control actuation devices. Fan-coil units and room air-conditioning units (both window unit and split-type unit) often employ a two-position control. In fact, room air-conditioning units usually adopt an on/off control, which is a special type of two-position control.

The operation of the 'two-position' control action can be illustrated by Figure 8.4. The difference between the 'on' and 'off' points of the thermostat is called the *control differential*. The difference between the high tempera- ture and low temperature in the room is called the *operating differential*. Generally speaking, the smaller the operating differential of the system, the better the system.

To improve the control action, most of the sophistication added to this simple two-position control is for the purpose of reducing the operating differential of the system. An example of methods to reduce the operating dif- ferential is to employ additional control actions, such as in the case of timed two-position control. Another example in HVAC systems is to employ a third control action: *floating action*. PID control can also be employed to improve performance of two-position control, namely two-position PID control or PID bang-bang control. It is worth noting that two-position control or on/ off control cannot control the process variable with high accuracy.

8.1.5 Temperature controls

Temperature control is the most fundamental function of air-conditioning systems. Temperature control is often done by a temperature controller called a thermostat which is set to the desired temperature value or set-point. The temperature deviation, or offset, from the set-point causes a control signal to be sent to the controlled device. P, PI or PID control functions are normally employed for temperature controls in systems with modulating valves or dampers.

In a chilled water or hot water coil, the temperature controller may modulate a water valve to vary the flow rate of the hot or chilled water through the coil. Air-handling units (AHUs) and primary air-handling units (PAUs) often employ modulating valves for precise temperature regulations nowadays. Figure 8.5 shows an AHU system with hot water heating and chilled water cooling. The room thermostat (*TC2*) controls the set-point of supply air. The temperature controller (*TC1*), using the set-point and measured temperature (*T1*), controls the heating coil valve (*V-2*) and cooling coil valve (*V-1*). In this case, sequential split-range control is usually employed for the supply air temperature control loop and cascade control is also often used for the system.

It is also common to use steam to heat air and direct expansion (DX) cooling to cool air. In the cold season, a temperature controller controls a steam valve to heat air. In the hot season, a temperature controller controls an expansion valve. The valves used to control the water flow through the coils may be either two-way or three-way (normally used in small-scale systems such as fan-coil units). In some systems, temperature can be controlled by controlling the dampers at the coil to vary the proportion of air passing through the coil, as shown in Figure 8.6.

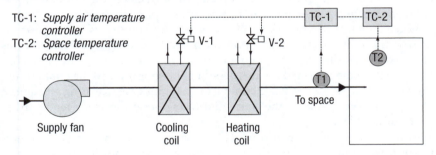

Figure 8.5 Schematic of temperature control of air-handling unit with water valves.

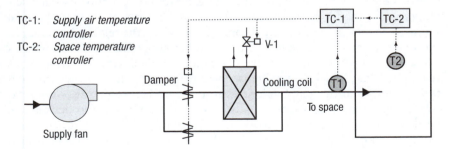

Figure 8.6 Schematic of temperature control of an air-handling unit with bypass damper.

8.1.6 Humidity control

Humidity control in a conditioned space is either for thermal comfort or industrial processes, which is done by controlling the amount of water vapour present in the air in the space. When relative humidity at the desired temperature set-point is too high, dehumidification is required to reduce the amount of water vapour in the air for humidity control. Similarly, when relative humidity at the desired temperature set-point is too low, humidification is required to increase the amount of water vapour in the air for humidity control. Commonly used dehumidification methods include:

- surface dehumidification on cooling coils simultaneous with sensible cooling;
- chilled water spray dehumidification with direct contact;
- direct dehumidification with desiccant-based dehumidifiers.

Humidification is not always required in HVAC systems, particularly in buildings basically requiring cooling. But it is usually required in buildings requiring heating, and is provided by a humidifier. Commonly used humidification methods include: *water spray humidification* and *steam spray humidification*. Dehumidification is usually done at the same time as the sensible cooling process is performed by a surface-cooling process on the cooling coils or direct-expansion refrigerant evaporator coils.

When cooling in an indoor space is controlled to remove water vapour for relative humidity control, the supply air might be cooled more than what is required for sensible or dry cooling of the space if humidity cannot be controlled within the expected range by cooling the supply air to the temperature just suitable to provide sensible cooling only. In this case reheating is required to prevent overcooling of the space and to maintain the space temperature at the desired level.

In many parts of the world which experience very dry winters, humidification is commonly needed. In a centralized air-conditioning system, a duct-mounted steam humidifier is usually used. A space humidity controller controls the indoor relative humidity at the desired set-point or within a certain range. A change in relative humidity from that set-point causes a control signal to be sent to the controlled component. When the relative humidity of the space drops below the humidity controller set-point, a control signal is generated to open the steam valve at the inlet to the duct-mounted humidifier unit. When the steam valve is positioned open, steam flows through the humidifier in the supply air stream to the space, which raises the relative humidity of the space eventually. A second humidity controller may be needed to be situated in ductwork downstream from the humidifier acting as a high-limit safety controller. When the relative humidity of the air stream approaches saturation point, the high-limit controller overrides the space humidity controller to reposition the steam valve and decrease the steam

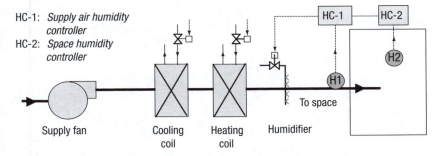

Figure 8.7 Schematic of humidity control of an air-handling system with steam spray.

flow. This may prevent condensation and water being carried downstream from the humidifier.

The arrangement of a cascade control can be used for space humidity control as illustrated in Figure 8.7. The space humidity controller (*HC-2*) sets the desired relative humidity set-point. When the space relative humidity drops below the humidity controller set-point, the controller will increase the supply air humidity set-point, and the supply air humidity controller (*HC-1*) will consequently open the steam valve to introduce more steam.

8.1.7 Static pressure control

In a variable air volume system, the supply air volume delivered varies with the building load. It is necessary to provide some means of fan capacity modulation and static pressure control in the duct system. Static pressure control will prevent excessive noise generation and energy loss as the variable air volume system reduces the demand of air volume. Damper throttling can offer a low-cost method of controlling air flow rate. However, it operates by adding resistance to the system or by 'destroying' fan pressure, and the amount of pressure is dissipated across the damper. In fact, the overall efficiency can be very low. The more effective means for energy savings is to change the fan speed or change the fan geometry (such as the pitch angle of axial fans). Multi-speed motors can be used. For example, a three-speed fan is commonly used in fan-coil units.

There are four typical types of fans with variable capacity usually used in HVAC applications:

1 variable-speed centrifugal fan;
2 axial fan with variable pitch angle;
3 fan with variable inlet vanes;
4 fan with variable outlet vanes.

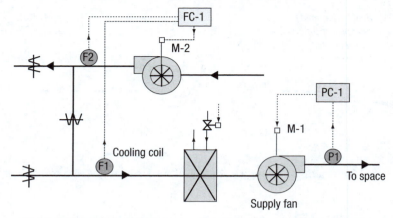

PC-1: *Static pressure controller*   M-1, M-2: *Driver or frequency inverter*
FC-1: *Return air flow controller*

Figure 8.8 Schematic of fan control of an air-handling system.

Comparison studies show that the partial load performance of types 1 and 2 fans are comparatively high. The partial load performance of type 3 fans is much lower and it is much worse for type 4 fans.

A typical fan control system of variable air volume in AHU systems is shown in Figure 8.8. The static pressure controller (*PC-1*) using the static pressure sensor (*P1*) controls the supply fan speed, pitch or vanes to maintain the static pressure at its set-point. The flow controller (*FC-1*) using flow rate sensors (*F1* and *F2*) controls the return fan to keep the flow difference between the supply and return fans so that they act as a team and maintain the static pressure of the space (positive or negative pressure as required).

8.2 Control of CAV systems

The air-handling unit is one of the most important components in HVAC systems. The processed air can provide heating or cooling, adjust the humidity and refresh the air for each conditioned zone. The control system must be able to optimize the temperature, humidity and fresh air ratio in the supply air to provide comfortable air with minimum energy consumption. The components of the air-handler include fans, heat exchanger coils, dampers, ducts, sensor and control instruments. The following operations may take place. *The outside air is admitted by an outside air damper. This air is then mixed with the return air from a return air damper. The mixed air is filtered, heated or cooled and humidified or dehumidified as required by the control system. Finally, the resultant supply air is delivered to the conditioned zones by various means.*

In Hong Kong, many AHUs, for both CAV and VAV air-conditioning systems, usually employ a centralized fresh air supply, i.e. primary air-handling unit (PAU), sometimes due to architectural constraints or the problem of outdoor air quality, and sometimes simply due to poor design practice. In this case, the flexibility of fresh air control is very limited. Usually, a constant-speed fan is used and no pressure control is needed. In many systems, the outdoor air intake cannot be controlled. In some other systems, the fresh air intake in AHUs can be controlled. The PAUs in these buildings may employ variable-speed fans. In these cases, the fan speed can be controlled to maintain the static pressure in the fresh air duct at a certain fixed set-point.

In CAV systems, almost no measure is taken to control fan operation, and the flow rate keeps nearly constant. Valves are controlled to regulate water flows in order to reach the desired supply air temperature. A typical control application of single-duct CAV systems is presented as follows.

8.2.1 Basic control of CAV systems

The system and its control are illustrated in Figure 8.9, which is a single-duct heating and cooling air-handling system with a return air fan and a supply air fan. The system is hot water or steam heating and chilled water cooling. It also uses an economizer control cycle to control an outdoor air (O.A.) damper. Usually outdoor air, return air (R.A.) and exhaust air (E.A.) dampers are interlocked. Generally, supply air leaving the cooling coil after surface dehumidification is often near a saturated condition and can meet indoor humidity requirements in cooling seasons. Therefore, no special measure is

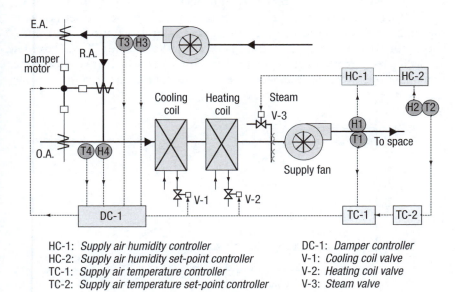

HC-1: *Supply air humidity controller*
HC-2: *Supply air humidity set-point controller*
TC-1: *Supply air temperature controller*
TC-2: *Supply air temperature set-point controller*

DC-1: *Damper controller*
V-1: *Cooling coil valve*
V-2: *Heating coil valve*
V-3: *Steam valve*

Figure 8.9 Control and instrumentation of a single-duct CAV system.

needed to control supply air humidity in cooling seasons. However, humidification is necessary in dry seasons, and humidification control can be used alone, which was discussed in Section 8.1. Humidity controller *HC-1* with supply air humidity sensor (*H1*) can regulate steam valve (*V-3*) according to the supply air humidity set-point set by the controller *HC-2* using space humidity sensor *H2*. The supply air temperature and fresh air control can be achieved with sequential split-range control. However, in many practical applications, the fresh air intake is set constant or controlled manually and the control is less complicated. In many applications, only cooling (coil) or heating (coil) is provided, and the situation is even easier to control.

The control of a situation with heating and cooling coils (without outdoor air control) is discussed below followed by discussion of the control strategy of the situation with a 'complete set-up'. It is important to notice that the sequential split-range control strategies for the control of heating and cooling coils and the control of heating and cooling coils plus outdoor air damper are the same as that used in VAV systems.

In the simplest case when only one cooling coil or heating coil is used, the control is simply a reconfigured cascade control. In the other two cases, the control is a combination of cascade control and sequential split-range control. The control strategy of the supply air temperature set-point reset controller (*TC-2*) is similar in all the CAV control cases. It can generate supply air temperature set-points using the measured space temperature (*T2*). In practical applications, PI or PID function is usually employed in the controller. Figure 8.10 illustrates the control strategy of the principle of a simple supply air temperature set-point reset controller. As the cooling load increases (or heating load reduces), the PI(D) output increases and the set-point of the supply air is reduced linearly, which is the simplest option and a good choice for control. The upper limit (*Tset$_{max}$*) and lower limit (*Tset$_{min}$*) are set according to system design or practical requirements, which are basically the limits that operators or designers have allowed, taking into account various

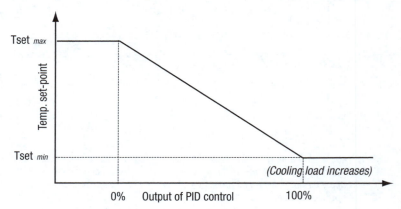

Figure 8.10 Control strategy of CAV supply air set-point reset controller.

issues. The selection does affect the control performance of the system if the correct operating range of the set-point is covered. For example, in pure cooling systems, the limits might be within the range between 12°C and 22°C. For pure heating systems, the limits might be between 22°C and 30°C. For systems with both heating and cooling, they might be between 12°C and 30°C. It is suggested not to give a too wide a range in applications in order to avoid unaccepted control, providing that the normal operating range of the set-point is covered.

8.2.2 Sequential split-range control of AHU

Sequential split-range control can avoid heating and cooling simultaneously in air-handling systems as shown in Figure 8.8. When the control employs outdoor air economizer control, the energy performance can be improved significantly as the result of maximum use of the free cooling capacity. Figure 8.11 shows the control strategy of sequential split-range control of the AHU temperature control loop in the situation where only heating and cooling coils are controlled sequentially.

Using the measured supply air temperature ($T1$) and set-point from the controller TC-2, the supply air temperature controller (TC-1) produces control signals for the heating and cooling coil valves. Figure 8.11(B) provides more detailed illustration of the relationship between the split-range control output (u) of the supply air temperature feedback controller (usually PI or PID control) and the control signals u_c and u_h (corresponding to cooling coil and heating coil) in the strategy. The controller output scale ranges from –100 to 100 per cent. The scale is somewhat arbitrary and is selected to emphasize the fact that each of the outputs to both actuators has a control signal corresponding to 0 to 100 per cent opening.

When the combined output of the feedback controls is between 0 and 100 per cent, it will be rescaled to 0 to 100 per cent to activate the cooling coil valve. Here, 0 per cent corresponds to the closed position of the cooling coil valve and 100 per cent corresponds to the fully open position, meaning maximum mechanical cooling. At this mode, the heating coil valve is closed.

When the combined output of the feedback controls is between –100 and 0 per cent, it will be scaled to 0 to 100 per cent to modulate the heating coil valve, and in the meantime the cooling coil is fully closed. Here, 0 per cent corresponds to the closed position of the heating coil valve, meaning no heating provided, and –100 per cent corresponds to the fully open position, meaning maximum heating is provided.

When outdoor air is controlled to achieve economizer control for maximizing the use of free cooling, the split-range AHU temperature control should involve control of the outdoor air damper. Similarly, using the measured supply air temperature ($T1$) and the temperature set-point, the supply air temperature controller (TC-1) produces control signals for the heating/cooling coil valves and the fresh air damper.

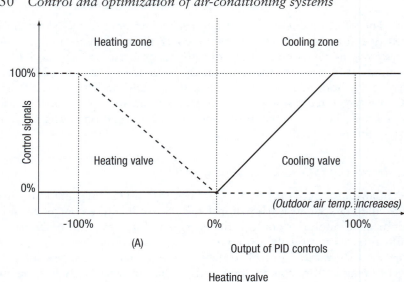

(A)   Output of PID controls

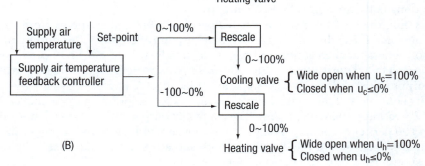

(B)

Figure 8.11 Sequential split-range control strategy for supply air temperature control of AHUs (outdoor air control optimization not included).

Figure 8.12 illustrates the split-range control strategy and Figure 8.13 provides more detailed description of the relationship between the split-range control output (u) of the supply air temperature feedback controllers and the control signals u_c, u_h and u_d (corresponding to fresh air damper control) in the strategy in this configuration. The combined controller output scale ranges from −100 to 200 per cent in the strategy as illustrated in the figures.

When the combined output of the feedback controls is between 100 and 200 per cent, it will be rescaled to 0 to 100 per cent to activate the cooling coil valve. Here 100 per cent corresponds to the closed position of the cooling coil valve, meaning no mechanical cooling, and 200 per cent corresponds to the fully open position, meaning maximum mechanical cooling. At this mode, the heating coil valve is closed. When the return air enthalpy is larger than or equal to the fresh air enthalpy, economizer cycle control (the fresh

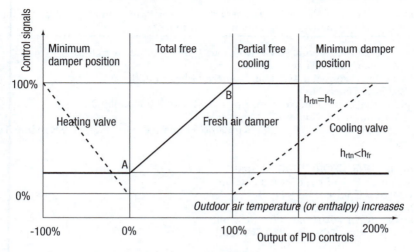

Figure 8.12 Split-range control strategy for supply air temperature control of AHUs including outdoor air control optimization.

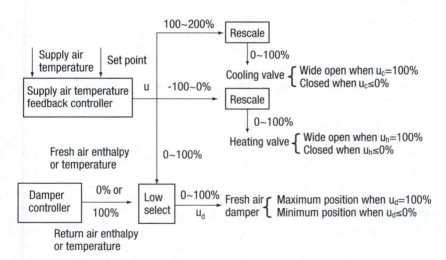

Figure 8.13 Relationship between split-range control output and control signals of fresh air damper and coil valves.

air damper fully open) is exploited simultaneously to minimize mechanical cooling demand, referred to as partial free cooling. When the return air enthalpy is less than the fresh air enthalpy, the fresh air damper will stay at its minimum position to control the fresh air flow rate at its lower limit.

When the combined output of the feedback controls is between 0 and

Table 8.1 An example of CAV temperature control strategy implemented in a high-level-language programming environment

---

Rpid = PID(Temp-room,TRset, Kp-room,TI-room,0.0,I0room,0.0,
  ER0room,TimeB,Time)
Hpid = PID(Temp-ahu,Tset-ahu, Kp-heat,TI-heat,0.0,I0heat,0.0,
  ER0heat,TimeB,Time)
Cpid = PID(Temp-ahu,Tset-ahu, Kp-cool,TI-cool,0.0,I0cool,0.0,
  ER0cool,TimeB,Time)
FApid=PID(Temp-ahu,Tset-ahu, Kp-damp,TI-damp,0.0,I0damp,0.0,ER0damp,Ti
  meB,Time)
TimeB = Time
Dtemp = RSCLTO(0.0,100.0,FApid)
Tset-ahu = RSCLFR(Tmax,min,Dtemp)
SUMpid = Cpid + Dpid – Hpid
Dheat = RSCLTO(0.0,-100.0,SUMpid)
Dcool = RSCLTO(100.0,200.0,SUMpid)
Ddamp = RSCLTO(0.0,100.0,SUMpid)
H-rtn = enthalpy(Temp-rtn,RH-rtn)
H-fresh = enthalpy(Temp-fresh,RH-fresh)
IF (H-fresh.LT.H-rtn) Ddamp = 0.0
OUT-heat = Dheat/100.0
OUT-cool = Dcool/100.0
OUT-damp = Ddamp/100.0

c   Parameters can (have to) be tuned:
c   Heating coil control: Kp-heat,TI-heat,   Cooling coil control: Kp-cool,TI-cool
c   Damper control: Kp-damp,TI-damp,   AHU temp reset: Kp-room,TI-room
c   Set-point:
c   TR-set: Room air temperature set-point
c   Tmin,Tmax: lower and upper limits of AHU temperature set-point
c   Measurement inputs:
c   Temp-room: room temperature
c   Temp-ahu: AHU supply air temperature
c   Temp-rtn: return air temperature
c   Temp-fresh: Outdoor air temperature
c   RH-rtn: return air relative humidity
c   RH-fresh: return air relative humidity

---

100 per cent, the fresh air damper will be modulated to control the ratio of fresh air and recycled air to ensure the supply air temperature is controlled at the set-point temperature (i.e. total free cooling is activated) by the cooling of outdoor air only. In this mode both heating and cooling coils are fully closed.

When the output from the feedback controller is between –100 and 0 per cent, it will be scaled to 0 to 100 per cent to modulate the heating coil valve and, in the meantime, the fresh air damper stays at its minimum position. Here, 0 per cent corresponds to the closed position of the heating coil valve, meaning no heating demand, and –100 per cent corresponds to the fully open position, meaning maximum heating demand.

An implementation example of a CAV system control including the AHU

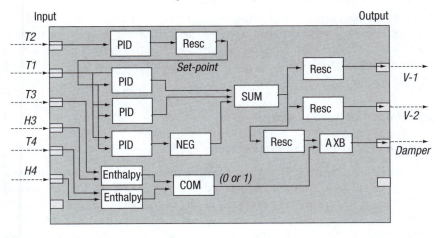

Figure 8.14 Example of CAV control strategy implemented in a symbolic programming environment.

Table 8.2 Function routines used in the control program listed in Table 8.1

```
     FUNCTION RSCLFR(E,F,G)
C Rescale from 0–100 to E-F
     RSCLFR=E+G*(F-E)/100.0
     IF (E.LT.F) THEN
     RSCLFR=MIN(F,MAX(E,RSCLFR))
     ELSE
     RSCLFR=MIN(E,MAX(F,RSCLFR))
     ENDIF
     RETURN
     END

     FUNCTION RSCLTO(E,F,G)
C Rescale to 0–100 from E-F
     RSCLTO=100.0*(G-E)/(F-E)
     RSCLTO=MIN(100.0,MAX(0.0,RSCLTO))
     RETURN
     END
```

split-range control (excluding outdoor air flow control), the room temperature
control and humidity control in a simplified symbolic programming environment for BAS stations is shown in Figure 8.14. An implementation of the
same strategy in a high-level-language programming environment is shown
in Table 8.1, where the Fortran language format is used. The PID function
presented in Table 7.4 is used. The function routines used, except the function for enthalpy calculation, are listed in Table 8.2. It is worth noting that
the programs presented here might have some stability problems due to the
interference of different PID functions and in the transient regions between

different modes, which will be discussed further in Section 8.4.4.When the strategy is more complicated in terms of programming, it is easier to pro-gram in a high-level-language programming environment. In a symbolic programming environment, programming is limited by the function boxes provided.

8.3 Control of VAV systems

8.3.1 Control of VAV air-handling units

Variable air volume systems represent one major type of HVAC system in use today. VAV systems are the results of developments of fans, motor drivers, air diffusers and control in response to global energy crises. VAV systems are divided into single-duct systems and dual-duct systems. In single-duct VAV systems, the supply air temperature is controlled to be constant or is reset by optimized supervisory control while the supply air volume changes to satisfy the cooling load. As the volume of the supply air to the zones through the terminal units (VAV boxes) changes, the air volume delivered by the fan must also be adjusted. The control methods employ mainly motor-actuated variable inlet vanes on the fan, variable-speed fan and variable pitch angle, which are described in detail in Section 8.1.7.

Figure 8.15 presents a control diagram of a typical single-duct VAV sys-tem with return and supply air fans equipped with variable speed control, although there are many variations in applications. Supply air temperature

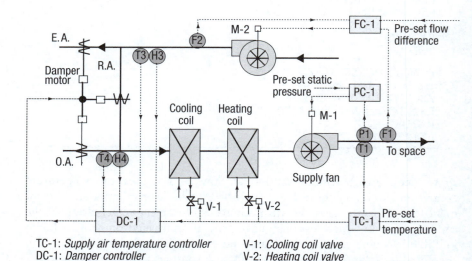

TC-1: *Supply air temperature controller*
DC-1: *Damper controller*
PC-1: *Supply air static pressure controller*
FC-1: *Return air flow rate controller*

V-1: *Cooling coil valve*
V-2: *Heating coil valve*
M-1: *Supply fan inlet vane damper motor*
M-2: *Return fan inlet vane damper motor*

Figure 8.15 Control diagram of a single-duct VAV system.

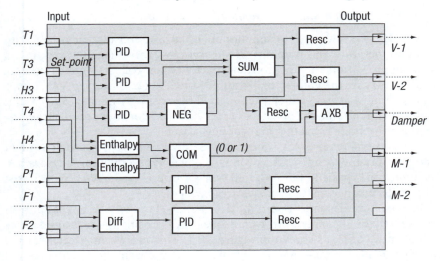

Figure 8.16 An example of control strategies of a VAV AHU system implemented in a symbolic programming environment.

control and economizer control are similar to that of a single-duct CAV system as discussed earlier. The static pressure in the main supply duct decreases or increases as terminal units control the supply air flow rate to meet the load requirements of different conditioned zones. The supply air static pressure controller (*PC-1*) controls the speed of motor (*M-1*) of the supply air fan to regulate the fan speed to keep the supply air static pressure at its set-point according to the difference between the measured static pressure *P1* and the static pressure set-point set by supervisory control (or manually set as constant). The return air fan is controlled by the flow controller (*FC-1*) to keep the flow difference between the supply air and return air or the ratio of return air to supply air at a set-point value to match the operation of the supply air fan. The inputs of the flow controller are flow rates of supply air and return air, and the pre-set flow difference or flow ratio.

An implementation example of the VAV system control including the AHU sequential split-range temperature control (including optimal outdoor air ventilation control), supply air fan (static pressure) control and return air fan control is shown in Figure 8.16 in a simplified symbolic programming environment for BAS stations.

The above system is designed to 'match' the supply and return air fans so that they act as a team and maintain space static pressure. At the same time, the supply air fan is modulated to maintain the supply air static pressure at the set point. In many applications, the return air fan is controlled to keep the space static pressure (positive or negative) as desired.

8.3.2 VAV terminal and room temperature control

The AHU of a VAV system supplies air usually at constant or nearly constant temperature and humidity. If supervisory control is employed, it can also supply variable temperature. Capacity is controlled to match cooling load by varying the volume of air supplied to a zone. Usually, a VAV box is provided in each zone which is connected to air supply duct(s). Such a unit can regulate the amount of air that enters the space or room to maintain the temperature both in heating and cooling seasons.

In terms of pressure, VAV boxes can be divided into pressure-dependent and pressure-independent. According to power types, they can be classified as induction types and fan-powered types. Fan-powered VAV boxes consist of parallel arrangement and series arrangement.

In a pressure-dependent unit, the damper *(D-1)* of the VAV box is controlled directly by the temperature controller *(TC-1)* using the measured space temperature *(T1)*, as shown in Figure 8.17. This means, in effect, that the pressure-dependent unit has no way of sensing change in the system pressure that may change due to a number of boxes throttling down and reducing the air flow of the system. These types of boxes depend entirely on the ability of the central fan system to adjust the flow as the boxes close off or open up. The result is that boxes near the fan get the majority of total air flow rate during heavy loads or pull-down periods. Moreover, seriously over-tuned or down-tuned phenomena should be very obvious because of the slow response of the building system as the change of pressure is sensed only after the room temperature is affected.

Figure 8.18 shows the control diagram of a pressure-independent VAV box. An example of the implementation of the control strategy in a simplified symbolic programming environment is shown in Figure 8.19. Temperature control is based on resetting flow set-point using cascade control to eliminate the effects of supply air pressure change on zone temperature control. In the

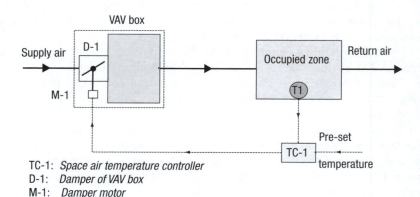

TC-1: *Space air temperature controller*
D-1: *Damper of VAV box*
M-1: *Damper motor*

Figure 8.17 Control diagram of a pressure-dependent VAV box.

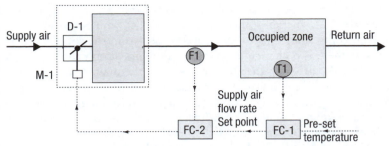

Figure 8.18 Control diagram of a pressure-independent VAV box.

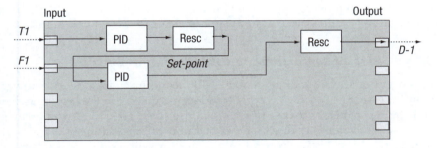

Figure 8.19 Example of pressure-independent VAV box control implemented in a symbolic programming environment.

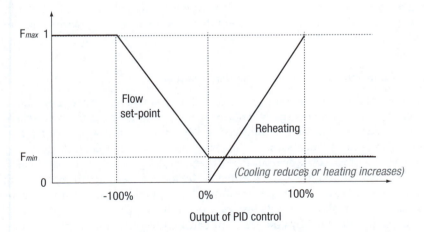

Figure 8.20 Schematic of temperature control strategy of a pressure-independent VAV box with reheating terminals.

unit, the damper (*D-1*) is controlled by the flow controller (*FC-2*) that is reset by the flow set-point controller (*FC-1*) according to the measured space temperature (*T1*). The set-point reset controller determines the flow set-point according to the change of cooling load sensed from the room temperature changes. If there is a change in the static pressure supplied to the VAV terminal (it varies, although the static pressure at the main duct is controlled), the flow rate will be affected and the change of flow rate is sensed by the flow sensor very quickly. The flow controller therefore regulates the damper to meet the required air flow rate. Eventually, the control system responds to the pressure fluctuations quickly before the room temperature is affected. The VAV box is therefore called a pressure-independent VAV box.

The pressure-independent unit is commonly used alone or with a reheating coil for reheating the air when additional heating is needed after the cooling is reduced. Usually, these types of boxes have a minimum amount of air that flows to a space, and that minimum amount of air is heated up by the reheating coil if it overcools an individual zone as shown in Figure 8.20. The reheating coils can be electric, hot water or steam.

8.4 Outdoor air ventilation control and optimization

A conventional air-handling control with economizer introduces a lower limit fresh air rate to keep acceptable indoor air quality in the heating mode and the mechanical cooling mode when the fresh air enthalpy is greater than the return air enthalpy, as described in detail in Sections 8.2 and 8.3. In fact, the conventional control results in over-ventilation by providing a constant or design ventilation flow rate when the occupancy ratio is low, and insufficient ventilation when the occupancy ratio is higher than the design or expected maximum occupancy. Consequently, it leads to energy waste and/or unsatisfactory indoor air quality. Demand-controlled ventilation (DCV) aims at achieving acceptable indoor air quality with minimum energy consumption.

8.4.1 Economizer control and its performance

As described in the previous sections (8.2 and 8.3), economizer control can be employed to save energy when it is beneficial to do so. In the cold season, economizer control can be employed to keep a minimum supply of fresh air. When temperature increases, cooling is required. Meanwhile, if the temperature (or enthalpy) is lower than that of return air, economizer control can be used to regulate the fresh air damper to introduce fresh air to maintain the supply air temperature at the set-point. When the fresh air damper reaches its maximum position with the increase of fresh air temperature and the cooling capacity provided by cool fresh air cannot meet the cooling requirement, mechanical cooling is necessary. If the fresh air temperature (or enthalpy) increases further and is greater than that of return air, the fresh air damper should remain at its minimum position to introduce only necessary fresh air to save the energy consumed by mechanical cooling.

The outdoor air economizer control can provide great benefit in saving energy in air-conditioned buildings located in cool, dry climates. In Hong Kong, it is usually considered that the use of an economizer is not beneficial and it is rarely adopted in Hong Kong practice. The design of most systems actually cannot allow the outdoor air flow rate to be controlled. However, economizer control can produce significant benefits even in Hong Kong. The following demonstrates the energy and environmental performance of economizer control used in Hong Kong (Yiu, Wang and Yik 2000).

The latitude of Hong Kong is 22°18', which is within the subtropical zone. Generally, the weather during January and February is relatively cloudy, with occasional cold but dry northerly winds. March and April have occasional spells of high humidity, and these months may be classified as the winter months.

Figure 8.21 shows the frequency distribution of the hourly outdoor air dry-bulb temperature, outdoor relative humidity and enthalpy in winter periods of five years (1994–9). Table 8.3 shows the activation frequency during the office hours in typical Hong Kong office premises. The occupation period was assumed as 08:00–18:00 h for weekdays, 08:00–13:00 h for Saturdays and not occupied on Sundays. The total activation frequency of the air economizer varied from 59.9 to 73.6 per cent in the office hours in winter in terms of different indoor air temperature set-points, which correspond to 20.0 to 24.5 per cent of total office hours. Considering the boundary conditions

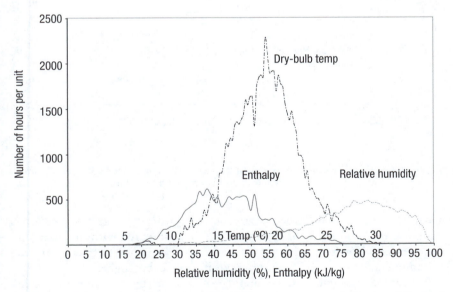

Figure 8.21 Distribution of outdoor dry-bulb temperature, relative humidity and enthalpy in five-year winter periods in Hong Kong.

Table 8.3 Activation frequency of economizer in office hours (9:00–18:00 h) in a
year in Hong Kong

Indoor set-point Temp (°C)/ Humidity (%)	Supply air temperature (°C)	Partial free cooling		Total free cooling		Free cooling overall	
		hour	% in winter	hour	% in winter	hour	% in winter
23.3/50	14	687	51.6	110	8.3	797	59.9
23.3/50	16	505	37.9	292	21.9	797	59.9
24/50	14	779	58.5	110	8.3	889	66.8
24/50	16	597	44.9	292	21.9	889	66.8
25/50	14	869	65.3	110	8.3	979	73.6
25/50	16	687	51.6	292	21.9	979	73.6

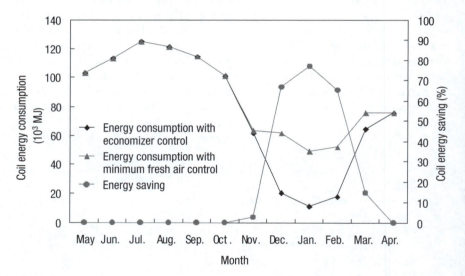

Figure 8.22 Monthly coil energy consumption with economizer control and minimum fresh air control and coil energy saving in Hong Kong.

when the enthalpy difference between the outdoor air and return air is
insignificant and the use of more outdoor air (compared with the minimum
requirement) benefits the indoor air quality, the activation frequency of fresh
air dampers for more outdoor air could be noticeably higher.

Figure 8.22 presents the percentage of coil energy consumption saving
and the monthly cooling coil energy consumption of the two AHUs in a year
using two different control strategies respectively (with economizer control

and with minimum fresh air control) in a Hong Kong commercial building from site monitoring. Compared to the case with minimum fresh air control, the monthly coil energy consumption from November to March was reduced by 3, 67, 77, 65, and 15 per cent respectively. The coil energy consumption when using the economizer cycle was reduced by 41.7 per cent in the winter period, which was estimated to be equivalent approximately to 12.1 per cent of the annual energy consumption of the cooling coil.

The site monitoring also shows that better indoor air quality was achieved when economizer control was used. Under partial free cooling modes, the outdoor air flow rate was set to maximum and more fresh air was drawn in to dilute the indoor pollutants. The mean and maximum CO_2 concentration in this case were found to be about 60–150 ppm lower than that recorded at the minimum fresh air mode. In total free cooling mode, CO_2 concentration was about 50–90 ppm lower than that at the minimum fresh air mode.

8.4.2 Demand-controlled ventilation

Outdoor ventilation airflow is one of the key factors affecting air quality in air-conditioned indoor spaces. DCV is one of the alternative strategies to achieve acceptable indoor air quality with minimum energy consumption. Current ASHRAE Standard 62–2004 requires that the minimum design outdoor ventilation air flow rate should be based on the occupancy of the space and the area of the space.

Measuring directly the actual concentration of pollutants in indoor air is ideally the best means of controlling indoor air quality. However, it is hard to find one sensor that is able to accurately sense all the major pollutants at the same time. Even for typical pollutants such as volatile organic compounds (VOCs), it is hard to find an accurate and reliable sensor that is reasonably simple and sufficiently low cost to allow it to be used in online control applications. Occupancy sensors are widely used in the control of lighting systems. They are suitable in cases when only the on/off (occupied or not occupied) status needs to be detected. Some other techniques, such as the computer vision technique, can be used to detect the number of occupants in a space. However, its application is obviously restricted by the geometrical layout of the indoor space for wide commercial applications in building management systems.

Use of CO_2 measurement for outdoor ventilation control is a well-developed technology and widely used method in ventilation-control applications where occupants are deemed to be the main source of pollutants. Although CO_2 in indoor spaces should be controlled within an allowable range, CO_2 control itself is generally not a great concern as it is normally well below the limit in air-conditioning applications. In the control system, CO_2 concentration is often used as the indicator of indoor pollutants as an indirect method as it is related to the degree of occupancy in the indoor space and therefore related to the pollutants generated by occupants. In other applications, such as car park ventilation, CO_2 cannot be used as a good indicator, while either CO or NO would be a better alternative.

The previous ASHRAE standard (62–1989) requested control of CO_2 concentration of the indoor space to within a limit of 1000 ppm. However, recent studies have demonstrated that this DCV strategy cannot adequately consider the ventilation demand in a space in many situations. First, the absolute level of CO_2 concentration is not a reliable indicator of ventilation demand. A space may be under-ventilated even when the CO_2 in the space is below a certain set limit. The air quality in a space may be acceptable, even when the CO_2 concentration in the space is over the same limit in other situations. Second, the DCV strategy cannot efficiently remove the non-occupant-generated pollutants (from building materials, furnishing, etc.) particularly when the pollutants generated by occupants do not dominate.

In the later version of the ventilation standard (62–2004), it is suggested that the minimum requirement for the outdoor air ventilation rate shall be determined not only by the actual occupancy but also the occupied area as shown in Equation (8–1):

$$DVR = R_p P + R_B A \qquad (8\text{–}1)$$

where DVR is the demanded (minimum) ventilation flow rate, P is the number of occupants, A is the area of the indoor space, R_p is the fresh air requirement per person, and R_B is the fresh air requirement per square metre. The approach provides a way to consider non-occupant-generated pollutants, but it requires identifying the actual occupancy in a space. A steady-state approach for detecting occupant numbers was proposed using CO_2 and flow measurements as shown in Equation (8–2), where E_{ac} is the ventilation effectiveness.

$$P = V_S E_{ac} (C_R - C_S)/S \qquad (8\text{–}2)$$

The CO_2-based DCV strategy might be achieved alternatively by controlling the differential CO_2 concentration of indoor space within the limit of 600–700 ppm above the outdoor CO_2 level of the building as stated in Standard 62–2004.

The studies of Wang and Jin (1998) show that there is significant delay in the steady-state method in responding to the changes of occupant load, particularly when the air change rate of the system is low. Therefore, Wang and Jin proposed a strategy based on the dynamic balance of the CO_2 in the occupied space as illustrated below. For a space served by a single-zone ventilation system as shown in Figure 8.23, or multiple-zone ventilation systems, the balance of CO_2 can represented by Equation (8–3):

$$PS + m_s C_s - m_s C_{rtn} = V \frac{dC_R}{dt} \qquad (8\text{–}3)$$

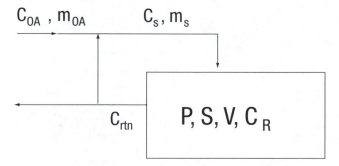

Figure 8.23 Balance of CO_2 in a single-zone ventilated space.

where m_s is the supply air volume flow rate, C_s and C_{rtn} are the CO_2 concentrations of the supply air and the return air respectively, V is the air volume in the conditioned indoor space, C_R is the CO_2 concentration in the conditioned space, and S the average CO_2 generation rate of an occupant.

On the basis of the CO_2 balance, two categories of method might be derived: the steady-state detection as shown in Equation (8–2) and dynamic detection. The dynamic-detection method considers both the steady-state CO_2 balance and the derivative (dC_R/dt) of the CO_2 concentration using the measured CO_2 concentration at the current and previous sampling instances as given in Equation (8–4):

$$\frac{dC_R}{dt} \approx \frac{C_R^i - C_R^{i-1}}{\Delta t} \tag{8–4}$$

where the superscripts i and $i-1$ represent the current and previous sampling instants respectively and Δt is the sampling interval. The occupancy (P^i) at the current sampling step is detected as Equation (8–5):

$$P^i = \frac{E_{ac}(m_{OA}^i + m_{OA}^{i-1})(C_R^i - C_{OA}^i)}{2S} + V\frac{C_R^i - C_R^{i-1}}{S\Delta t} \tag{8–5}$$

On the right-hand side of the equation, the first part considers the steady-state balance and the second part considers the change rate of the CO_2 concentration, where m_{OA} is the fresh air volume, C_{OA} is the CO_2 concentration of the fresh air and E_{ac} is the air change effectiveness.

8.4.3 Sequential split-range control of AHU combining DCV control

DCV aims at providing minimum but sufficient outdoor air flow rate to buildings in order to minimize the mechanical cooling load from outdoor ventilation. Certainly, outdoor air introduced to an indoor space will generally increase the level of indoor air quality. However, it is not always best to

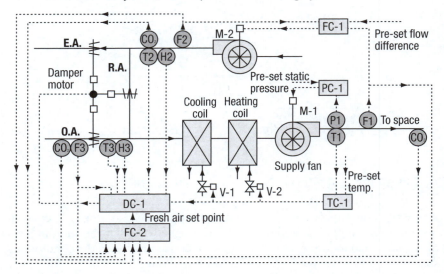

TC-1: *Supply air temperature controller*   V-1: *Cooling coil valve*
DC-1: *Fresh air damper controller*   V-2: *Heating coil valve*
PC-1: *Supply air static pressure controller*   M-1: *Supply fan inlet vane damper motor*
FC-1: *Return air flowrate controller*   M-2: *Return fan inlet vane damper motor*
FC-2: *DCV based fresh air set point controller*

Figure 8.24 Control diagram of VAV system with DCV control.

minimize the outdoor air intake. Outdoor air economizer control optimizes the ventilation control by concerning itself with the energy issue only. To minimize the energy use while providing sufficient or minimum required ventilation, the best approach is the combination of economizer control and demand-controlled ventilation.

Figure 8.24 shows a schematic diagram of an outdoor air ventilation control strategy of a VAV system combining economizer control and DCV control. Temperature control, return fan control and supply fan control are the same as that shown in Figure 8.15. The main difference is that DCV control is introduced and combined with the economizer control. The temperature controller *TC-1* generates the control signals for modulating the heating coil valve, cooling coil valve and the fresh air damper (in total free cooling mode) to control the supply air temperature. DCV-based fresh air set-point controller *FC-2* generates an outdoor air flow rate set-point (demanded minimum outdoor air flow rate) for the damper controller *DC-1* to maintain acceptable indoor air quality when DCV is beneficial, which determines the flow rate set-point according to the actual occupancy load. The supply air temperature and fresh air control are performed by sequential split-range control.

Figure 8.25 illustrates the sequential split-range control strategy combining DCV control. When DCV control is preferable, damper control (*DC-1*)

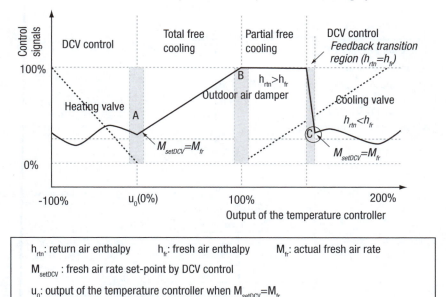

Figure 8.25 Sequential split-control strategy of VAV system with DCV control.

exploits the actual fresh air flow rate (*F3*), with respect to the set-point fresh air flow rate from the fresh air set-point controller (*FC-2*), to generate the signal for the fresh air damper control. In total free cooling mode, the fresh air damper controller (*DC-1*) receives a control signal from the supply air temperature controller (*TC-1*) to control damper position to maintain a desired supply air temperature.

Figure 8.26 presents a detailed description of the relationship between the combined PI(D) feedback control output u and the control signals u_c, u_b and u_{dT} (for the cooling coil valve, the heating coil valve and the fresh air damper respectively), and the relationship between the temperature control loop and the DCV based fresh air damper control loop. The temperature control output scale ranges from -100 to 200 per cent. When DCV control is preferable, fresh air damper control output is between 0 and 100 per cent.

When the output for the temperature control loop is between 100 and 200 per cent, it will be rescaled to 0 to 100 per cent to actuate the cooling coil valve to cool the air (where 100 per cent corresponds to the closed position of the cooling coil valve, and 200 per cent corresponds to the fully open position). In the meantime, according to the return air enthalpy and the fresh air enthalpy, economizer logic is utilized to determine whether the DCV-based fresh air damper control is activated. When the return air enthalpy is larger than or equal to the fresh air enthalpy, economizer cycle control (the fresh air damper is fully open) is exploited simultaneously to reduce the mechanical

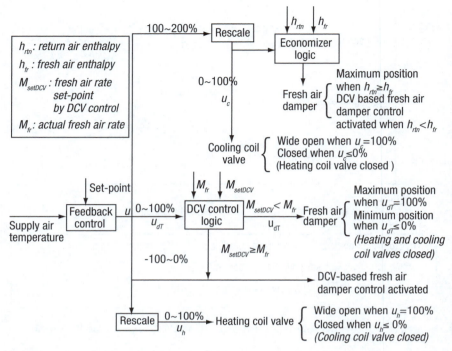

Figure 8.26 Logic of AHU sequential split-range strategy combining DCV control.

cooling demand, namely partial free cooling. When the fresh air enthalpy is larger than the return air enthalpy, DCV control is used and the fresh air damper is regulated according to the control signal from the DCV-based fresh air damper control, u_{dDCV}, to keep acceptable indoor air quality.

When the output from the temperature control loop is between 0 and 100 per cent, the fresh air damper is adjusted to control the amount of the fresh air flow to ensure the supply air temperature remains at the set-point (i.e. total free cooling is activated). In this process, the DCV control logic is used to determine whether the fresh air flow based on temperature control is sufficient or whether the DCV-based fresh air damper control should be applied. If the set-point of the fresh air flow based on DCV control is larger than or equal to the actual fresh air flow rate, the DCV-based fresh air damper control is activated to modulate the fresh air damper to take in sufficient fresh air. Simultaneously, the output from the feedback temperature control is reset to 0 per cent by resetting the PID terms of the three temperature-based control loops.

When the output from the temperature control loop is between 0 and −100 per cent, it is scaled to 0 to 100 per cent to modulate the heating coil valve to heat the air (i.e. the process enters heating mode). Here 0 per cent represents the closed position of the heating coil valve (i.e. no heating demand), and

−100 per cent represents fully open position (i.e. maximum heating demand). In this process, the DCV-based fresh air damper control is activated.

8.4.4 Stability of AHU sequential split-range strategies

In the simplest case of split-range control as shown in Figure 8.11, there exists the problem of instability when the controller changes its mode between heating and cooling coil controls due to the dead zone or rangeability of valves, and the controllability when the valve is near closed position. When the characteristics of heating and cooling coils are different, the control parameters suitable for two processes are different. In the split-range strategies presented in this chapter, multiple PIDs are used, which allow PIDs for different processes to be tuned individually. However, as the PIDs can be activated simultaneously even though only one is desired, the PIDs of other processes affect the control of the running process.

When the control of the outdoor air damper is involved in the AHU sequential split-range control, as shown in Figures 8.12 and 8.13, the problem is more obvious as the characteristics of the damper can be much more different from that of coils and more changeover regions are involved. When the AHU split-range control involves DCV control, the situation is more complicated and the problem can be even more serious.

When AHU systems utilize economizer control to reduce energy consumption, control difficulties often occur in the transient region between heating and total free cooling, and at the transient region between total free cooling and partial free cooling. Such control instability results in waste of energy by using alternation modes and has a negative impact on comfort conditions. In addition, it causes unexpected wear and tear on valves, dampers and actuators.

A finite state machine (FSM) sequencing control strategy was introduced by Seem *et al.* (1999) for AHUs to overcome the difficulties. Simply speaking, the FSM strategy adopts certain delays in two transient regions, one between heating mode and total free cooling mode, and the other between total free cooling mode and partial free cooling mode, and assumes a dead band temperature in the transient region between partial free cooling mode and mechanical cooling mode plus minimum fresh air flow rate, when shifting the control from one mode to another.

However, the problem of interference between PIDs for different processes remains and the fixed control parameters cannot produce good performance for the entire operating conditions. Gain scheduling is a technique that deals with nonlinear processes, processes with time variation or situations where the requirements on the control change with the operating conditions.

Generally, in the control strategy for AHU with economizer control, the supply air temperature is utilized, with respect to the set-point temperature, to determine control signals to the heating coil valve, the fresh air damper and the cooling coil valve. When DCV control is adopted, the fresh air rate set-point determined by DCV control is exploited in respect of the actual fresh air rate to produce the control signal for the fresh air damper. Evidently, there are extra control difficulties due to the fresh air damper control, which lead to instability phenomena such as alternation and oscillation in the transient region between DCV control plus heating mode and total free cooling mode as well as in the transition process from partial free cooling mode to DCV control plus total mechanical mode, as shown in Figure 8.25. A more robust control strategy including three robust schemes at these three different transient regions respectively is proposed by Wang and Xu (2002) to overcome these control instability problems, which includes:

- A robust transition control scheme with I-term reset and gain scheduling in the transient region between DCV mode and total free cooling mode.
- A 'freezing' transition control scheme with gain scheduling in the transition process between total free cooling mode and partial free cooling mode.
- A feedback transition control scheme with I-term reset in the transition from partial free cooling mode to DCV mode.

By 'freezing' the PIDs of the processes which are not in operation, the strategy allows only one PID to be activated at one time and the interference among PIDs is avoided. Gain-scheduling technique is employed in the transient regions (when a valve/damper is near closed position or near fully open position) to improve control stability. The I-term reset scheme allows a smooth start of a process control at the beginning when it takes over the control duty by setting a proper initial value of its I-term.

8.5 An overview of optimal control methods used for HVAC systems

An overall classification of main optimal (or supervisory) control methods used in HVAC systems is illustrated in Figure 8.27. Supervisory control in HVAC systems can be classified into four categories, including model-based optimal control method, hybrid optimal control method, performance map-based optimal control method, and model-free optimal control method. Such

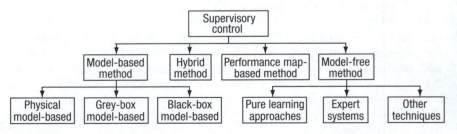

Figure 8.27 Classification of optimal control methods in HVAC systems.

a classification may not be ideal since there are no clear boundaries between some control methods. However, it can provide a very useful and helpful basis for comparing the advantages and disadvantages of different control methods. It is also very helpful for identifying the strengths and weaknesses of each method as well as for analyzing the feasibilities of their online applications. The control methods using physical models, grey-box models and black-box models can be classified into the category of model-based methods while the methods using expert systems and pure learning approaches can be grouped into the model-free category.

8.5.1 Model-free optimal control methods

Model-free optimal control methods do not require any 'model' of the targeted system. Expert systems and the reinforcement learning approach can be utilized to design the model-free optimal control methods. An expert system acts as an optimal controller, as it has the capability to determine the most energy-efficient or cost-efficient control settings for the optimal operation of HVAC systems according to a given working condition. These energy- or cost-efficient control settings are identified based on the combination of the rules defined in the knowledge base and information obtained from the BASs. The knowledge base in an expert system is derived from the specific knowledge of one or more human experts. An expert system can imitate human reasoning to make decisions for given working conditions based on the knowledge base. It also has the ability to deduce reasonable solutions with an incomplete set of data. An expert system is easy to program and easy to manage. However, application of expert systems is strongly affected by the richness of the knowledge database since the rules are static and anything outside its domain of expertise may result in significant errors.

Reinforcement learning control is another example of a model-free optimal control method. This method describes a learning paradigm in which a control system attempts to improve its behaviour on the results of previous actions without the requirement of a model of the environment or the effects of actions. This method can find the optimal or near-optimal solutions for the control problem without any prior knowledge of the environment. However, it always takes an unacceptably long time to make the controller 'learn'. The performance of the controller is sensitive to many factors.

8.5.2 Model-based optimal control methods

In model-based optimal control, the tools required to perform the optimal control are the system and/or component models and optimization techniques. The main function of the models is to predict the system energy or cost and environment performance as well as the system response to the changes of control settings. All the models are related to power consumption and/or operating cost directly. Online measurements collected from BASs are used to tune the model parameters to make them represent the actual system. The primary role of the optimization technique is to seek the energy-efficient and/or cost-efficient control settings (i.e. operation mode and set-points) to minimize the system energy input or operating cost while still maintaining satisfactory controlled variables. At a sampling instant, the optimization technique is applied to these models to evaluate the control settings that minimize power consumption and/or operating cost as characterized by the models. The control strategies determined in this manner react quickly to the rapid changes of internal and external conditions. According to the knowledge of the system utilized to formulate the models, the model-based optimal control can be further divided into physical model-based optimal control, grey-box model-based optimal control and black-box model-based optimal control.

8.5.3 Hybrid optimal control methods

In hybrid optimal control, different types of models and/or model-based control method and model-free control method are combined together to formulate the optimal control strategies. For instance, some hybrid optimal control methods utilize a mix of physical/grey-box/black-box models to design the control system, in which some component models are physical models while others are grey-box or black-box models. Some hybrid optimal control methods use both the model-based approach and model-free approach (e.g. reinforcement learning approach) to construct the optimal control methods, in which the features of model-based approach and model-free approach are combined to achieve high control performance. The optimal control methods formulated in this manner may provide good control performance if the controllers are designed reasonably well.

8.5.4 Performance map-based optimal control methods

Compared to the three methods presented above, performance map-based optimal control is somewhat different. This method often uses results generated from the detailed simulation (or experimental tests) of the targeted system over the range of expected operating conditions to draw a performance map, and then utilizes this map for optimal control of HVAC systems. For instance, various combinations of cooling loads, ambient air temperatures, the number of operating chillers, the number of operating pumps as well as the number of operating cooling towers and their individual fan speeds can be used as inputs to the simulation platform based on detailed component models of an electrically driven chiller plant without significant thermal energy storage. At each operating condition, the power consumption or performance data for all combinations are computed, and the control settings giving minimum energy value or best performance are identified. A performance map can then be drawn using those combinations with minimum energy values or best performance identified from over the full operating range of a system, and in addition can be used as a optimal controller for optimal operation of the HVAC system.

8.6 Optimal control of air-side systems

To save energy consumption and improve an indoor environment, various optimal control set-points, such as DCV-based fresh air set-point (which was addressed in detail in Section 8.4), supply air temperature set-point reset, VAV supply air static pressure set-point reset and so on can be used for air-handling systems. In a typical VAV system, the local loop controls and the supervisory (or optimal) control are listed as follows (Wang 1999, Wang and Jin 2000).

Local loop controls include:

1   AHU supply air temperature control;
2   static pressure control;
3   zone temperature control (zone air flow control);
4   outdoor air flow rate control;
5   return fan control.

Supervisory (optimal) controls include:
1   outdoor air flow rate set-point reset;
2   static pressure set-point reset;
3   AHU supply air temperature set-point reset.

The local loop controls are essential for the system to operate properly to provide a comfortable and healthy indoor environment, while the supervisory controls allow the system to operate with a better or optimized performance. The outdoor air flow rate set-point reset affects the energy consumption (mainly coil load) and the indoor air quality. The static pressure reset affects the energy consumption of the fan only. However, this resetting may affect the indoor environment if it does go to an incorrect range. If it is set too low, the critical zone may not be provided with sufficient air flow. If it is set too high, noise problems may occur. The AHU supply air temperature set-point reset will affect the thermal comfort and energy use of the system (mainly fan consumption).

8.6.1 Online optimal reset of static pressure set-point

The control strategy aims at minimizing the VAV fan energy consumption by minimizing static pressure. In order to supply sufficient air to every individual zone as well as to minimize static pressure, the static pressure is controlled to be just sufficient for the most heavily loaded zones. The online strategy makes use of all the VAV damper positions represented by relevant VAV damper position control demands as the indicator of relative load of the individual VAV terminals as shown in Figure 8.28.

The local DDC control signals can be conveniently collected by the supervisory controller when the AHU and VAV control stations are integrated in a single network. The static pressure is adjusted just allowing that the VAV damper with the highest relative cooling load among all the VAV terminals is controlled to be very close to fully open position at any time, in order to ensure that all the individual zones are supplied with sufficient air and the static pressure is controlled at its lowest allowable level.

When a large number of VAV terminals are associated with one AHU, a certain limited number of VAV dampers may be allowed to be fully open (or over the threshold) by setting the parameter (N_{max}). Therefore, the first N_{max} largest damper positions are neglected and the ($N_{max}+1$)st is used as the maximum control damper position by the controller. When the maximum damper position is over a threshold, which is set as the set-point of the PID, one PID function is activated to increase the static pressure set-point. By contrast, the other PID function is activated to reduce the static pressure set-point when the maximum damper position is below the threshold. A hysteresis is added as a 'dead band' of set-points to increase the stability of the control strategy. The static pressure set-point is determined by rescaling the difference between the outputs of two PID functions within a pre-set pressure range.

In some applications, the summations of the demanded flow set-points and the actual measured flow rates of VAV boxes are used as the indicator. If the summation of actual measured flow rates of VAV boxes is less than that of the set-points, the pressure set-point is not high enough to deliver sufficient

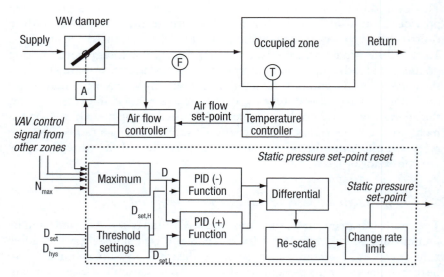

Figure 8.28 Schematic of VAV static pressure set-point reset strategy.

air flow to the air-conditioned zones. In cases where pneumatic controllers are used for VAV boxes or the VAV controllers are not integrated with the AHU controllers, it is not possible for the AHU controller to have the information on the control of VAV boxes, so some 'near optimal' control could be adopted. The optimal set-point can be set as a function of the total supply air flow rate according to experience or site monitoring or site tests. When the total supply air flow rate reduces, the pressure set-point can be reduced.

8.6.2 Supply air temperature set-point reset

The air flow rate to a zone is reduced by closing down the VAV damper when the load of a zone is low. Under low partial load, the total flow rate of a zone may be reduced to be very low in order to meet the reduced load. The reduction in total flow rate results in significant savings in fan power. On the other hand, a low ventilation flow rate may cause deficiencies in system performance (e.g. poor mixing of supply air and room air, inadequate room ambient air circulation and dumping). When a minimum limit of the total flow rate is used, a space may be overcooled under low partial load if the supply air temperature is low.

A proper resetting of the supply air temperature allows the VAV system to avoid poor ventilation and save fan power. A strategy for online reset of the supply air temperature set-point is illustrated in Figure 8.29. The strategy utilizes the air flow rate set-points of the pressure-independent VAV box

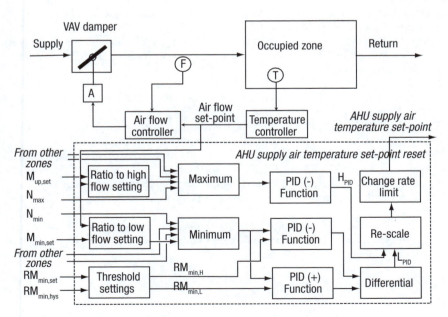

Figure 8.29 Schematic of AHU supply air temperature set-point reset strategy.

controllers as the cooling load indicators of individual zones. The minimum flow set-point is set to avoid performance deficiency in individual zones, which may have different values for different zones and need to be selected and tuned according to the actual design and situation of individual zones. The upper flow rate set-point is a parameter used only for calculating the relative load of a zone. Since different zones may have very different flow ranges, the air flow rate set-point of each zone is normalized by using the ratios of the set-point to the minimum flow set-point and upper flow set-point respectively.

The maximum ratio to upper flow set-point among zones is selected as the indicator of the relative load of the zone with the most critical thermal load. The minimum ratio to minimum flow set-point among zones indicates the most critical zone in terms of ventilation. A positive value of L_{PID} indicates that flow in a certain zone is too low and the supply air temperature set-point needs to be increased. A large value of H_{PID} indicates that the load of a certain zone is high and the supply air temperature set-point needs to be reduced. The outputs of the PID functions are rescaled within a pre-set range to determine the temperature set-point. The change rate limit is applied to the temperature set-point to ensure the stability of system control.

References

CIBSE. (2000) *CIBSE Guide H – building control systems*. Oxford: Butterworth-Heinemann.

Seem, J. E. Park, C. and House, J. M. (1999) 'A new sequencing control strategy for air-handling units', *HVAC&R Research*, 5(1): 35–58.

Wang, S. W (1999) 'Dynamic simulation of building VAV air-conditioning system and evaluation of EMCS on-line control strategies', *Building and Environment*, V34(6), pp. 681–705.

Wang, S. W. and Jin, X. Q. (2000) 'Model-based optimal control of VAV air-conditioning system using genetic algorithm', *Building Environment*, 35(6): 471–8.

Wang, S. W. and Jin, X. Q. (1998) 'CO_2-based occupancy detection for on-line outdoor air flow control', *Indoor and Built Environment*, V7(3), pp. 165–81.

Wang, S. W. and Ma, Z. J. (2008) 'Supervisory and optimal control of building HVAC systems: a review', *HVAC&R Research*, 14(1): 3–32.

Wang, S. W. and Xu, X. H. (2002) 'A robust control strategy for combining DCV control with economizer control', *Energy Conversion and Management*, 43(18): 2569–88.

Yiu, C. M., Wang, S. W and Yik, W. H. (2000) 'Assessment on practical applications of outdoor air economizer in Hong Kong', *Building Services Engineering Research and Technology*, 21(3): 187–98.

9 Control and optimization of central chilling systems

Central chilling systems consume about 25 to 50 per cent of annual energy budgets in most air-conditioned commercial buildings. Well-monitored and controlled central chilling systems have great potential to improve operation reliability and reduce overall energy consumption. This chapter presents the energy characteristics and basic control of chillers, typical chilling system configurations, the general approaches for optimal control of chilling systems, the optimal control of heat-rejection systems and the systematic optimization of heat-rejection systems, the optimal set-point reset of chilled water temperature, the sequence control of multiple chiller plants and the optimal control of chilled water distribution systems.

9.1 Basic knowledge of chillers

9.1.1 Basic components and typical types

Generally, a central chilling system consists of three subsystems:

- *Chillers*, whose function in a central chilling system is to produce an adequate quantity of chilled water at the required temperature.
- *Heat-rejection system*, whose function is to reject the heat from the refrigerant to the environment.
- *Chilled water distribution system*, whose function is to distribute the chilled water to the user terminals.

In a central chilling system, the heat (cooling load) is carried by the chilled water to the evaporator, where the heat is transferred to the refrigerant. The refrigerant takes the heat to the condenser where it is passed on to the cooling water.

A typical mechanical chilling system involves at least one compressor (reciprocating, rotary-screw or centrifugal) and two heat exchangers (the evaporator and the condenser) as well as a heat-rejection system such as a cooling tower. The system operating cost is the cost of circulating the fluids (refrigerants, water, etc.) in the heat exchangers and the compressor.

Four types of chillers are normally used in centralized chilling systems: reciprocating chillers, rotary-screw chillers, centrifugal chillers and absorption chillers, where the first three types belong to the category of vapour compression chillers, which are discussed in this chapter. The energy-saving potential of central chilling systems depends on operating conditions, space constraints and the willingness of the building owners to pay for energy-saving features and measures.

9.1.2 Basic working principles

A vapour compression chiller removes an amount of heat (Q_e) from the chilled water and invests an amount of work (W) to deliver a quantity of heat (Q_c) to the environment.

In the ideal pressure-enthalpy cycle of the chiller, there are two isothermal and two adiabatic processes as shown in Figure 9.1. As the refrigerant goes through the expansion valve, the high pressure sub-cooled refrigerant liquid becomes the liquid-vapour mixture at a low pressure and low temperature. As it goes through the evaporator, this becomes a superheated low-pressure vapour stream. Then, in the compressor, the pressure and the temperature of the refrigerant vapour are increased. Finally, this high-temperature vapour is condensed at a high pressure. The adiabatic processes take place in the expansion valve and the compressor. The isothermal processes take place in the evaporator and the condenser. The isothermal processes in this cycle are also isobaric since they occur at constant pressures. The refrigerant carries the heat from a low to a high temperature level.

The coefficient of performance (*COP*) of a chiller is defined as the ratio between the heat removed from the process (Q_e) and the work (W) required to achieve this heat removal, as shown in Equation (9–1). The COP is

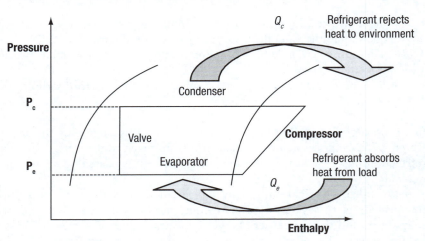

Figure 9.1 Schematic of the ideal refrigeration cycle.

normally greater than 100 per cent. Typically, chillers operate with COPs within the range of 2.5 to 7.0. Optimization can increase the COP by increasing the evaporating temperature or decreasing the condensing temperature of the refrigerant.

$$COP = \frac{Q_e}{W} = \frac{Q_e}{Q_c - Q_e} \qquad (9\text{--}1)$$

9.2 Chiller capacity control and safety interlocks

The cooling capacity of a chiller is the maximum refrigeration effect which can be delivered to the designated building process under particular operating conditions. The cooling capacity is practically never constant. The purpose of the capacity control is to make sure that the cooling capacity delivered by the plant can match the actual cooling load, making chillers operate energy efficiently.

The conventional approach of adjusting capacity to cooling load, especially in a chiller system with multiple reciprocating compressors, is by simply stopping a compressor in case the capacity exceeds the cooling load and starting an additional compressor in case the cooling load exceeds the capacity of the operating compressors. Nevertheless, under part load conditions, the increase of the switching frequency of the plant is not desirable since the switching actions adversely affect secondary process variables and also shorten the compressor's service life.

The capacity control of chillers can basically be accomplished in five different ways, including:

1   thermostatic expansion valve;
2   inlet guide vane;
3   hot gas bypass;
4   variable speed driver;
5   slide valve.

Among them, the variable speed driver control has the highest energy efficiency and the hot gas bypass control is the lowest. The inlet guide vane and slide valve are two capacity controls unique to centrifugal chillers and rotary-screw chillers respectively, and the rest of the controls can be applied to all vapour compression chillers.

An interlock system prevents the compressor motor from being started if any of the following conditions exist and stops the compressor if any of the following (except the first condition) occurs:

- open suction vane, detected by limit switch;
- low evaporator water temperature, near freezing point, sensed by low temperature switch;

- low water flow, sensed by low-flow switch;
- high compressor discharge pressure, sensed by high-pressure switch;
- high motor bearing or winding temperature, detected by high-temperature switch;
- low oil pressure.

Another interlock system guarantees that the following components are operating upon starting the compressor: water pump, oil pump and water to oil cooler. The suction vane usually has an interlock to be sure that it is completely closed when the compressor stops.

9.3 Chillers and central chilling system configurations

9.3.1 Chillers using different heat-rejection methods

According to the actual environmental constraints of the buildings, three methods are normally adopted by chillers for their condenser heat rejection:

- *water-cooled system;*
- *air-cooled system;*
- *cooling tower system.*

When a site is near a sea or a lake, the sea water or lake water can be used. The limited fresh water resources and government restrictions on the use of cooling towers in the past have resulted in Hong Kong using sea water cooling as one of the popular heat-rejection methods for chilling systems in buildings. Direct sea water cooling and indirect sea water cooling are two options. In indirect sea water-cooled systems, the sea water cools the condenser indirectly by passing through a sea water heat exchanger where the condenser cooling water is cooled. In direct sea water-cooling systems, the sea water cools the condenser directly by passing through the pipelines of the condenser. In this case, efficiency is generally high as an intermediate heat exchanger is omitted. However, the condenser itself should be made of special materials.

Air-cooled heat-rejection systems are used widely in small and middle-sized chiller systems and heat pump systems, especially in transportation refrigeration systems. In Hong Kong, it has been the normal system utilized when a building is remote from the seafront. An air-cooled condenser is an air-to-refrigerant heat exchanger, where air absorbs the heat from the refrigerant directly without any mass transfer between them. The air-cooled condenser is preferred in areas where the available water source is limited.

Usually, fans are used to enforce the heat transfer between the air and the refrigerant in the condenser. The air flow rate through the condenser affects chiller performance while the air temperature is not controllable. The increase of air flow rate (fan speed) will reduce the condensing temperature of chillers and therefore reduce the power consumption of chillers.

The cooling tower, a very conventional rejection system, is an air-to-water heat exchanger. In a cooling tower, the heat and mass transfer processes are combined with cooling water, which absorbs the heat from the refrigerant and then rejects the heat to the air. The mass transfer from evaporation consumes water. This loss is about 5 per cent of the water needed for an equivalent once-through cooling by a stream of water. Cooling towers are capable of cooling water until 2.5 to 5.5 K above the ambient wet-bulb temperature. The larger the cooling tower for a given set of water and air flow rates, the smaller this temperature difference will be.

9.3.2 An overview of chilled water systems

The goal of a chilled water system is to produce and distribute chilled water with high reliability, low capital cost and high flexibility. The central chilling system can be divided into two loops: the primary system which is also called the production system – the place where chilled water is produced; and the secondary system which is also called the distribution system because its function is to convey the chilled water to the load side. Typically, there are three pumping configurations applied to chilled water systems: *constant primary-only pumping system*; *constant primary/variable secondary pumping system*; and *variable primary-only pumping system*. For all options, the distribution system is variable flow using two-way valves or constant flow using three-way valves at cooling coils. It is well known that the use of three-way valves wastes energy because of the flow in the distribution system being greater than necessary in low-load conditions. The distribution pumping system (secondary pumps for constant primary-secondary variable system and primary pumps for variable primary system) can be fitted with variable-speed driven pumps or constant-speed pumps.

9.3.3 Constant primary-only pumping system

This is usually a constant differential pressure-control system. The schematic diagram of this pumping system for a typical application is illustrated in Figure 9.2. In this pumping system, a set of pumps serves the chillers and the load side. Two-way valves are used on the load side and pumps are operated at a constant speed. The differential pressure bypass valve (DPV) helps in maintaining a constant chilled water flow through the chillers, which is an essential condition for steady operation of the chillers.

The controlled variable in this control system is the differential pressure between the main supply and return pipes. A bypass pipe is installed to link up the two pipes across which the differential pressure is monitored and controlled. At the bypass pipe, a flow control valve is installed and this valve is modulated to open or close, under the control of the differential pressure controller. When the terminal cooling load decreases, the chilled water flow rate through the cooling coils will drop because of the modulating of the two-

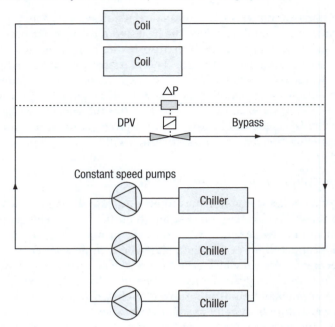

Figure 9.2 Schematic of a constant primary-only pumping system with differential pressure bypass valve.

way valves. Consequently, the differential pressure will exceed the set-point level, and the controller will command the DPV to open wider so as to allow more chilled water flow through the bypass pipe, thus the differential pressure is relieved. The reverse happens when there is a drop in the differential pressure until the flow control valve is fully closed. This constant primary-only pumping system with a DPV control is a very common chilled water system in constant water flow systems. To achieve higher energy efficiency, variable water systems are commonly adopted, reducing the pump power at off-design load.

9.3.4 Constant primary/variable secondary pumping system

For many years, the design of chilled water plants has been dominated by the constant primary/variable secondary paradigm. The schematic diagrams of this type of pumping system are shown in Figure 9.3. The rationale for this system is straightforward. On the secondary system (distribution system), constant chilled water flow wastes large quantities of pumping energy at part load. On the primary system (production system), however, chiller manufacturers recommend that the flow through the evaporator of a chiller should be constant for steady control and safety.

The primary constant speed pumps require less power than the secondary pumps because they only need to overcome the friction losses associated within the chillers, pipes and valves in the primary loop. The primary pumps are balanced against the design water flow rate. The secondary pumps require higher power because they must overcome the friction losses associated with the entire secondary loop (i.e. pipes, valves, coils etc.). Both constant-speed pumps and variable-speed pumps are normally used in secondary loops.

When the flow demand reduces from m_1 to m_2, if constant pumps are used, the pump characteristic curve remains unchanged and the system characteristic curve changes as the valves in the terminal units are closed down to achieve the new balance as shown in Figure 9.4A. If variable-speed pumps are used in the secondary loop, both the pump characteristic curve and the system characteristic curve can be changed, depending on the control implemented, by slowing down the pumps and closing down the valves in the terminal units. Different new balances can be achieved as illustrated in Figure 9.4B. The balance *Point D* is the worst case (same as in the constant-pump system) when energy efficiency is of concern. The balance *Point B* is the best case when the new balance is achieved by slowing down the pump speed only.

The main benefits of the constant primary/variable secondary water system are its simplicity and fail-safe nature besides saving pump energy at partial load. In many installations, the bypass pipe is simply a water pipe, allowing the chilled water to pass in both directions as shown in Figure 9.3A. In many more recent installations, a check valve is installed in the bypass pipe as shown in Figure 9.3B. With degrading ΔT (the difference between chilled water return and supply temperatures) due to low load or degraded cooling coils, the check valve in the bypass pipe can allow chillers to be fully loaded before sequencing on an additional chiller and this can reduce the system energy consumption. Moreover, the check valve can prevent water flow in

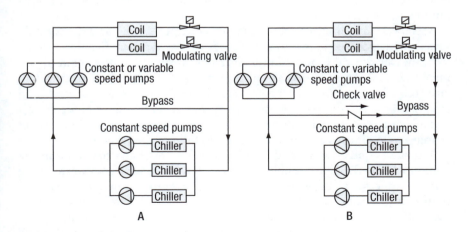

Figure 9.3 Configurations of constant primary/variable secondary pumping systems.

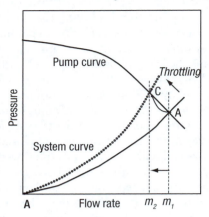 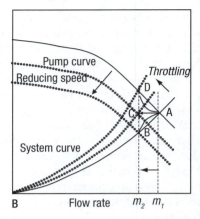

Figure 9.4 Control and balance of the secondary chilled water loop when the water flow rate demand changes from m_1 to m_2: A. Constant-speed pump system; B. Variable-speed pump system.

the reverse direction, that is, mixing of the primary supply and secondary return water, which dilutes the chilled water supply temperature in the secondary system.

9.3.5 Variable primary-only pumping system

Recently, there has been notable criticism of the conventional constant primary/variable secondary pumping system. Many researchers argue that it is impossible to optimize the energy consumption of a central chilling plant without adopting variable primary flow (VPF) systems which minimize conditions of excess flow. Figure 9.5 shows the schematic diagram of this pumping system.

As shown in Figure 9.5, the variable-speed pumps in the primary-only system are controlled to maintain the differential pressure at the most remote location (the critical branch) in the distribution system at a set-point determined to be sufficient to deliver the required chilled water flow rate through any coil. The set-point may be constant or reset downwards at part-load conditions. The variable-speed pumps can significantly reduce pump energy and are cost effective in most variable-flow chilled water applications, where changes in load demand require a variable chilled water flow rate.

A bypass control valve and a flow meter are installed to maintain the minimum flow through the chillers, due to the fact that the flow rate in the primary system may be below the minimum flow rate required by the chiller evaporators when there is very low demand for chilled water flow rate from the coils. The flow rate is sensed by the precise flow meter. The controller then opens the bypass valve to maintain the required minimum flow through the chillers.

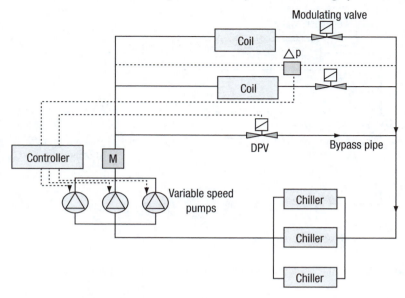

Figure 9.5 Configuration of variable primary-only pumping system.

This system has two significant disadvantages: the complexity of the bypass valve control and the complexity of the chiller sequencing control. Both result in a high risk of system failure unless careful engineering considerations are taken into account.

9.4 Chiller performance and optimal control

9.4.1 Chiller energy performance

The energy consumed by a chiller itself is basically used to transfer the refrigerant from low pressure (evaporating temperature) to high pressure (condensing temperature) as illustrated in Figure 9.6. The condensing and evaporating temperatures establish the refrigerant pressure and thus the energy needed per unit cooling load (see Figure 9.7).

Considering the efficiency of a chiller alone, the condensing temperature should be maintained as low as possible and the evaporating temperature should be maintained as high as possible to minimize refrigerant head which means less compressor energy consumption. Theoretical improvements of chiller COP by raising evaporating temperature and lowering condensing temperature for typical refrigerants under a typical air-conditioning situation are illustrated in Table 9.1. A decrease of 1 K in condensing temperature or an increase of 1 K in evaporating temperature can result in an increase of about 3 per cent in chiller COP.

In reality, the actual effect is influenced by various losses, and the saving

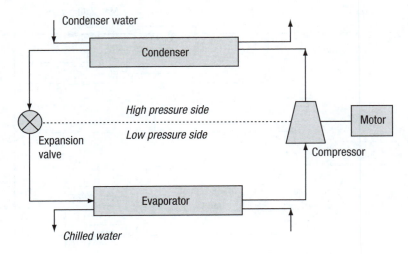

Figure 9.6 Schematic diagram of a water-cooled chiller system.

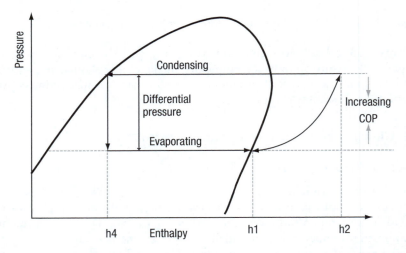

Figure 9.7 Pressure-enthalpy diagram for an ideal refrigeration cycle.

is significantly less than that obtained from the theoretical calculation. Normally, a saving of 1.5–2.5 per cent can be expected by reducing the condensing temperature by 1 K or increasing the evaporating temperature by 1 K in practical chillers at normal air-conditioning conditions. There is also a minimum condensing temperature limit for chillers which varies due to different chiller designs. The evaporating temperature should be as high as possible, which is determined by the specific requirements and the working conditions of the air-conditioning terminal units.

Table 9.1 Theoretical improvements of chiller COP by raising the evaporating temperature and lowering the condensing temperature

Refrigerant	R134a	R12	R22	R502	R717
Condensing temp. = 40°C Evaporating temp. = 0°C	4.336	4.486	4.453	3.978	4.802
Condensing temp. = 38°C Evaporating temp. = 0°C	4.636	4.786	4.747	4.274	5.093
Condensing temp. = 40°C Evaporating temp. = 2°C	4.616	4.773	4.728	4.234	5.106
COP *rise per degree reduction in condensing temp. (%/K)*	3.46	3.34	3.30	3.72	3.03
COP *rise per degree increase in evaporating temp. (%/K)*	3.23	3.20	3.09	3.22	3.17

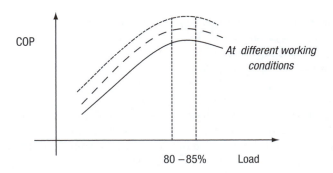

Figure 9.8 Relations between the cooling load and chiller COP.

The load of a chiller also strongly affects its efficiency, which is mainly due to the efficiency of the compressor at different loads. Generally, the efficiency of a chiller with one compressor at lower load is lower than that at higher load with the same condenser and evaporator working conditions. However, for a typical centrifugal chiller, this rule is a little different. The fact is that a centrifugal chiller usually achieves its maximum COP when it is running within the load range of 80 to 85 per cent, which is demonstrated in Figure 9.8.

9.4.2 An overview of optimal control of central chilling systems

As introduced above, the central chilling system of a building consists of three subsystems: chillers, heat-rejection system and chilled water distribution system. All three subsystems consume energy. The three main subsystems of a chilling system interact with each other and the fact is that the reduction of power consumption in one system would normally result in the increase of power consumption in the other systems. Furthermore, the increase of power consumption in one subsystem is normally different from the reduction of power consumption in the other subsystems when the operating conditions change. Therefore, the systematic optimization is not to minimize the power consumption of one system, but to minimize the power consumption of the entire system.

The whole chilling system must be operated as a coordinated single process. The basic idea of the systematic optimization is to minimize the sum of the power consumption of the three subsystems while fulfilling certain demanded cooling loads at certain external conditions as illustrated by Equation (9–2):

$$W_{tot}\big|_{\min} = (W_{p,cb} + W_{chil} + W_{cd})\big|_{\min} \qquad (9\text{–}2)$$

where

W_{tot}: *total power consumption of the entire chilling system*;
$W_{p,cb}$: *power consumption of chilled water pumps*;
W_{chil}: *power consumption of chillers*;
W_{cd}: *power consumption of the condenser heat-rejection system*.

The power consumption of each subsystem is determined by the load and operating condition. That means that the items on the right side of Equation (9–2) are the functions of the loads and operating conditions. In practice, one more difficulty is that the performance of the subsystems (i.e. the functions in the above equation) may change due to performance degradations as the results of ageing, fouling and so on. Therefore, the tasks or steps of optimal control of chilling systems include the following items:

1   Identify the load and relevant external environment.
2   Identify or update the performance of chiller subsystems.
3   Determine the optimal operating settings of the system including the subsystems, which lead to minimum total energy consumption of the entire system.
4   Control the system and subsystems at the determined optimal settings.

The second step is optional in practical applications while it is essential when the degradation of the system or some subsystems is of concern. In practical control applications, accurate and reliable identification of the load and performance of subsystems is the most difficult task.

In a particular installation, the task of the systematic optimization is to find the optimal cooling water temperature from cooling towers (cooling tower systems) or the optimal condenser cooling air flow rate (air-cooled systems) or the optimal sea water pump speed (sea water-cooled systems) and the

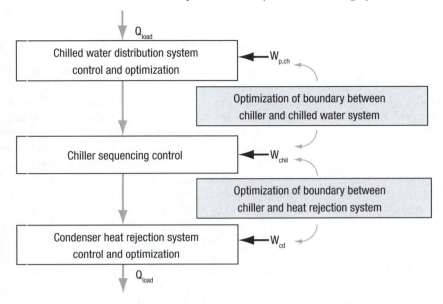

Figure 9.9 Five categories of optimal control of central chilling systems.

optimal chilled water supply temperature that will result in meeting the cooling needs of the installation at overall minimum energy consumption. After these optimal set-points are determined, variables are controlled to maintain the whole chilling system with these optimal set-points.

The important fact is that the optimal points vary due to the change of cooling loads and working conditions, such as outdoor air temperature and sea water temperature as well as the performance degradation of subsystems. All those make the online optimal control a more difficult task.

The functions of optimal control of central chilling systems may be grouped into five categories as shown in Figure 9.9, including the control and optimization of three subsystems (heat-rejection system, chillers and chilled water distribution system) and the systematic optimizations between three subsystems (i.e. between chillers and chilled water distribution system, and between chillers and heat-rejection systems). The optimizations of the three subsystems are local optimization. The optimizations on the control sharing of the efforts among three subsystems are considered as systematic optimizations. The systematic optimization between chillers and chilled water distribution system is, in fact, the optimal set-point reset of the chilled water supply temperature.

9.5 Optimal control of heat-rejection systems

9.5.1 Optimization of sea water-cooled systems

Since sea water-cooled systems use sea water to cool the condenser water (as shown in Figure 9.10), the temperature of the sea water cannot be controlled. The balance between the power consumption of chillers and pumps should be optimized since the high sea water flow rate reduces the chiller power consumption by lowering the chiller condenser water supply temperature while requiring more pump power. To facilitate optimization, variable speed pumps are introduced into condenser heat-rejection systems, providing significant energy-saving potential. However, investigations show that the energy-saving potential of variable-speed pumps has not been fully realized in many buildings. For instance, the saving of the pump power consumption is often compensated by the increase of the chiller power consumption. Suitable online control strategies are essential for optimizing the operation of variable-speed pumping systems in order to realize actual energy savings.

Correct reset of the sea water differential pressure control set-point can provide up to a 10 per cent saving in total chilling system energy consumption, while a 5 per cent saving can be expected from most cases investigated (Wang and Burnett 2001).

9.5.2 Optimization of air-cooled systems

Besides constant-speed fans, multiple-speed fans and variable-speed fans are introduced for cooling the air-cooled condensers to improve the overall chiller efficiency by varying the air flow rate at partial load. The air flow rate through the condenser affects the chiller performance while the ambient air temperature is not controllable. The increase of the air flow rate (fan speed) will reduce the condensing temperature of chillers and therefore reduce the power consumption of chillers. However, the increase of air flow rate will increase the power consumption of condenser fans at the same time. The

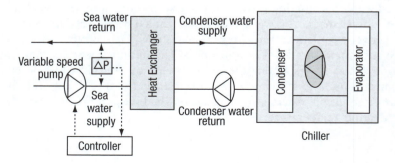

Figure 9.10 Sea water-cooled heat-rejection system.

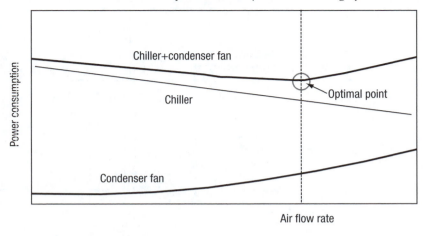

Figure 9.11 Power consumption vs. air flow rate in an air-cooled system.

optimal point of fan speeds to be controlled is that which allows the total power consumption of chillers and condenser fans at the minimum at a certain load and ambient air temperature. Figure 9.11 illustrates the relations between the air flow rate and power consumption in air-cooled systems with variable-speed fans.

9.5.3 Cooling tower optimization

Control of cooling tower supply water temperature: To minimize chiller energy use, the condenser water supply temperature set-point (i.e. the cooling tower supply water temperature) should be as low as possible. However, the control set-point should be at or above the lowest temperature attainable by cooling towers at a certain air (wet-bulb) temperature to avoid the waste of fan energy trying to reach an unobtainable value. As the difference between these temperatures increases, the cost of operating the cooling tower fans drops and the costs of pumping and compressor operations increase.

The cooling tower fan speed control is the usual method of maintaining the pre-set cooling tower supply water temperature (as shown in Figure 9.12). In cool weather or partial load, a tower bypass valve can be used to further reduce the cooling capacity when the fan speed is at its low limit. Besides the application of variable speed fans which can increase cooling tower energy efficiency at partial load, the cooling tower fans may be multiple (two or three) speed fans and centrifugal units with variable blade pitch which can provide continuous throttling.

Aims of cooling tower supply water temperature optimization: Optimizing the cooling tower supply water temperature is an important part of chiller optimization. A decrease in the cooling tower water supply temperature tends to reduce the cost of chiller operation but it increases the amount of

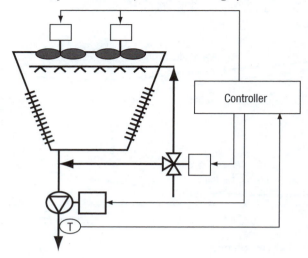

Figure 9.12 Control of the cooling tower supply water temperature.

work of cooling tower fans. Each 1 K reduction in the cooling tower supply water temperature can reduce the operating costs of the chiller compressor by about 2.5 per cent.

The optimal cooling tower supply water temperature is the temperature that can satisfy the load at the total minimum cost of all the operating equipment, such as chillers plus cooling tower fans and pumps. The relations between the cooling tower supply water temperature and the power consumption are shown in Figure 9.13. This optimal value is a function of the load and the weather conditions (wet-bulb temperature). Traditional control systems have usually operated with a constant cooling tower supply water temperature of 24°C or higher.

Optimal and near-optimal controls of cooling tower supply water temperature: Three strategies are introduced here including: *fixed approach, near optimal control strategy*, and *optimal strategy*. The fixed approach control method varies the cooling tower air flow rate to maintain a constant temperature difference (named approach) between the cooling tower supply water temperature and the ambient air wet-bulb temperature. Usually, the design approach (i.e. for most cases this is 5 K) is used. In this strategy, the ambient air wet-bulb temperature needs to be continuously monitored to derive the cooling tower supply water temperature set-point.

For real-time control and operation of cooling towers, near optimal control strategies can also be used. An example is the empirical model proposed by Sun and Reddy (2005) as shown in Equation (9–3), to determine a near optimal condenser water supply temperature set-point for cooling tower control, which is a function of the ambient air wet-bulb temperature and the load ratio of chillers, where, h_0, h_1 and h_2 are coefficients. The near optimal

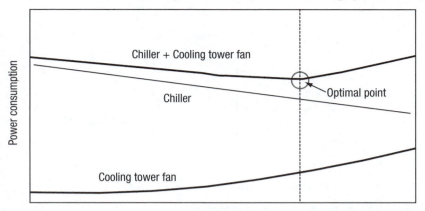

Figure 9.13 Power consumption vs. the cooling tower supply water temperature in a cooling tower heat-rejection system.

control strategies are commonly generated from performance maps while performance maps may probably be generated from complete simulation and/or site investigation.

$$T_{w,cd,\text{sup}} = h_0 = h_1 T_{wb} + h_2 (Q_{ev}/Q_{ev,des}) \tag{9–3}$$

These basic and near optimal strategies are simple enough and easy to implement in practice. However, they cannot provide the true optimal settings, which might provide settings significantly different from the optimal values, and a significant amount of energy might still be wasted. It is also worth noting that these strategies can only provide the condenser water supply temperature set-point. They cannot provide the best operating combination of the number of cooling towers in operation and their individual fan speeds that can control the system to operate at the desired temperature set-point.

To maximize the operating efficiency of cooling towers, model-based optimal control strategies can be used. In the model-based control strategy, two basic issues are performance models and optimization tools. The performance models are used to predict the system energy performance and the system response to the changes of control settings, while optimization tools are used to solve the optimization problem and identify the most energy-efficient control settings. For online applications, simplified models could be the best choice. The optimization tool used should be robust for identifying the global optimal settings while still requiring manageable computation time.

For practical applications, a hybrid quick search (HQS) method has been proposed by Ma, Wang and Xiao (2009) as an optimization tool for model-based control strategies for cooling towers. Using the HQS optimization tool, the near optimal control setting determined by the near optimal control

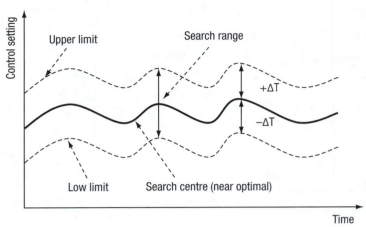

Figure 9.14 Schematic of the defined search ranges based on the near optimal settings.

strategy, for instance, Equation (9–3), is used as the search centre to define a relatively narrow search range, as illustrated in Figure 9.14 for the control variable to be optimized. This search range can be mathematically defined as in Equation (9–4):

$$T_{w,cd,\text{sup}}^{n,o} - \Delta T \leq T_{w,cd,\text{sup}} \leq T_{w,cd,\text{sup}}^{n,o} + \Delta T \tag{9–4}$$

Based on this narrow defined search range, the right, simple optimization tool, such as the exhaustive search method, can then be adopted to seek the global optimal solution for the given condition within this range and with a proper increment. The optimization technique determined by this manner can find the global optimal solutions reliably, while still satisfying the requirements and constraints of practical applications. A comparison study between an HQS-based optimal control strategy and a GA-based optimal control strategy showed that the HQS-based optimal control strategy can find the global optimal solutions as can the GA-based control strategy, but the computational cost was reduced greatly (by 96 per cent).

Cooling tower sequence rules: Usually, multiple cooling towers are employed as the heat-rejection system in a large chilling system. Correct sequencing of their operation is an essential function in order to achieve stable operation and lengthen equipment life. The sequencing rules are as follows:

- All variable-speed fans should operate at equal speeds.
- For multiple-speed fans, when adding cooling tower capacity, the speeds of the lowest speed fans are increased. When reducing cooling tower capacity, the speeds of the highest speed fans are reduced.

The main idea of these rules is to maintain the towers at the same load as far as possible. The above rules are also applicable to the control of multiple fans of air-cooled condensers and these rules may also be extended to the control of parallel pumps in the heat-rejection system.

9.6 Optimal set-point reset of chilled water supply temperature

9.6.1 Variable water volume systems

The systematic optimization of chilled water distribution systems aims at minimizing the overall power consumption of chillers and chilled water pumps while fulfilling the demand for cooling from user terminals. When the evaporator temperature rises as a result of controlled chilled water supply temperature from chillers increasing, the power consumption of chiller compressors is reduced as the suction pressure rises. However, the power consumption of chilled water pumps depends on the temperature difference across the user terminals and the temperature difference is mainly affected by the chilled water supply temperature. The higher the supply temperature, the more water that needs to be pumped to transport the demanded amount of heat from user terminals and the more power that is consumed by the water distribution pumps.

When the return temperature is constant, each 1 K increase of the supply temperature reduces the compressor operating cost by about 2–3 per cent and increases the chilled water pump operating cost by about 10 per cent. It gives some idea of the magnitude of the energy effects, although the operations in practice, particularly in variable water volume systems, are much

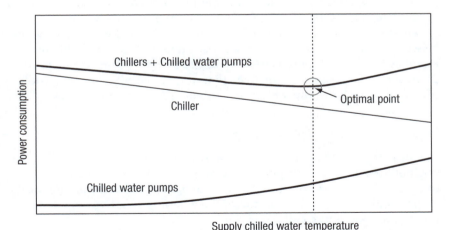

Figure 9.15 Power consumption vs. chilled water supply temperature in variable water volume systems.

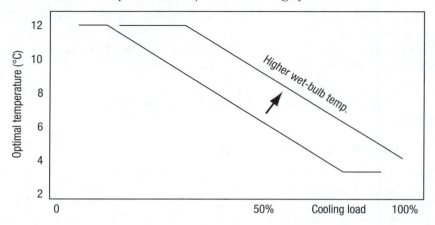

Figure 9.16 Near optimal chilled water supply temperature vs. building cooling load in chilling systems with variable-speed chilled water pumps.

more sophisticated. Figure 9.15 shows the relations between the chilled water supply temperature and the power consumption in variable water volume systems. In variable water volume systems, the optimal set-point is the best temperature set-point at a certain fixed setting of a chilled water loop. An optimal set-point exists at a certain load and working condition as illustrated by Figure 9.15, which changes as the load or working condition changes.

As optimizing the chilled water supply temperature set-point in systems with variable speed chilled water pumps is not an easy job in practical applications, some simplified methods have been proposed. For example, Braun (ASHRAE 2007) proposed a near-optimal control for the chilled water supply temperature set-point as illustrated in Figure 9.16. The control is based on the observation that the optimal set-point of a building's central chilling system decreases as the total building cooling load increases and the relation between them is approximately linear. Such numerical relation is different for different buildings and needs to be identified by site-monitoring or tests for particular buildings. It may be noted that, for chilling systems using cooling towers, the line moves up when the wet-bulb temperature increases.

9.6.2 Constant water volume systems

In constant water volume systems, the situation is much simpler. The set-point of the chilled water supply temperature should be set as high as possible to satisfy the cooling needs of all user terminals as the consumption of the pumps is almost constant and the optimization is simply to minimize the power consumption of chillers.

A cooling load following control can be used to determine the optimal chilled water supply temperature, which can be rather easily implemented in chilling systems whose BASs are integrated with the digital control stations

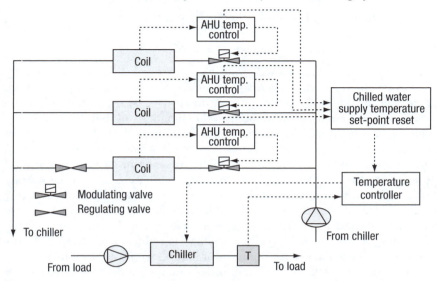

Figure 9.17 Load following control of the chilled water supply temperature.

of the terminal units. Figure 9.17 shows a system for continuously measuring and maintaining the optimal value for chilled water supply temperature. The controller monitors the water control valve positions of representative AHUs to reset the chilled water supply temperature set-point until one or more AHU valves are fully open and their AHU discharge air temperatures can reach their set-point. The following simple rules can inform such a control algorithm:

- If all water valves are unsaturated or the discharge air temperatures of all AHUs with saturated valves are lower than the set-point, increase the chilled water supply temperature set-point.
- If more than one valve is saturated at 100 per cent open and their corresponding discharge air temperatures are greater than their set-points, decrease the chilled water supply temperature set-point.

Note that the optimal chilled water supply temperature set-point achieved by the above cooling load following control is, in principle, the optimal setting for constant water volume systems and the systems with constant chilled water pumps only. In these cases, the objective of optimization is to save the power consumption of chillers as much as possible since the power consumption of pumps is constant or almost constant. However, it is not really the optimal set-point for variable water volume systems with variable chilled water pumps as different settings can be reached if the differential pressure of the chilled water loop is set at different values.

9.7 Sequence control of multiple chiller plants

9.7.1 An overview of chiller sequence control

Centralized chilling systems with multiple centrifugal chillers are commonly used in commercial buildings for providing cooling to the occupied spaces. Over one-third of total energy consumption is used by these chillers in most air-conditioned commercial buildings in a climate like Hong Kong's. Large buildings normally employ multiple numbers of chillers. There are many advantages of using multiple chillers in a building's central chilling system. The running chillers can remain at high efficiency in partial system load as some chillers can be turned off and the running chillers can be heavily loaded. Therefore, operation of chillers in extremely low load can be avoided, where a surge of centrifugal compressors may happen or chillers may lose temperature control. The life of equipment is lengthened as the total number of running hours of an individual machine is reduced, resulting in lower maintenance cost and reduced chances of failure. Some of the chillers can be set aside for routine maintenance in order to improve the overall reliability of the system.

When multiple chillers are employed, sequencing their operation correctly becomes an essential function, which must be provided by the control systems for central chilling systems. Such control is normally called chiller sequence control. The chiller sequence control provides economical loading and unloading of chillers. The control can be as simple as manually turning on and off a second chiller. However, the ultimate aim of the sequence control is to make the operating chillers achieve an overall COP as high as possible while fulfilling the demanded cooling load. A proper chiller sequence plays an important role in the overall performance of an air-conditioning system. It can become very complex when multiple chillers of different cooling capacities, chiller types and efficiency are installed.

There are various methods of chiller sequence control employed in various buildings with different complexities in terms of control parameters and equipment. Their differences mainly lie in how the instantaneous building load is measured and the ways to determine how many and which chillers are to be put into operation accordingly. It should be mentioned that details of the chiller sequence control are also dependent on the part-load characteristics of the chillers and their associated chilled water circulation pumps as well as on the circuit design of the chilled water distribution system. However, this section focuses on the sequence control methods and their performance. Typical chiller sequence control methods include:

- *temperature-based sequence control;*
- *bypass flow-based sequence control;*
- *direct power-based sequence control;*
- *total cooling load-based sequence control.*

9.7.2 Temperature-based sequence control

As mentioned earlier, the chilled water distribution network is often divided into two circulation loops: the primary loop and the secondary loop. There is a set of chilled water pumps in each loop and the two loops are hydraulically decoupled by a bypass pipe (Figure 9.18). The pumps ensure that the particular differential pressure in each loop is sufficient to deliver the required water flow.

For temperature-based sequence control, a water temperature sensor is installed at the main chilled water return pipe, downstream of the bypass line, which transmits the temperature signal to the sequence controller. The chilled water supply temperature is controlled at a constant set-point by the chiller internal control. When the chilled water return temperature exceeds the upper limit of the control band, an additional chiller will be sequenced online. When the chilled water return temperature drops blow the lower limit of the control band, the running chillers will be switched off, one by one. When the chilled water return temperature stays within the control band, no further change to the status of on and off stages will occur.

In chiller systems, the chilled water pumps are normally switched on and off together with their interlocked chillers so that the chilled water flow rate in the central chilled water plant varies in steps according to the number of chillers online. The temperature-based sequence control relies on the proper operation of the differential pressure bypass control which makes sure that the chilled water return temperature is a reasonable indicator of

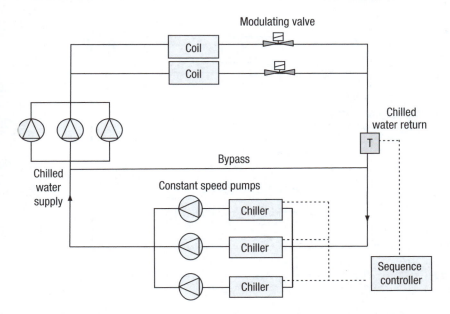

Figure 9.18 Chiller sequence   control based on the chilled water return temperature.

the simultaneous cooling demand load. However, the situation in practical operation is very complex, and chiller sequence control based solely on the chilled water return temperature cannot provide precise and reliable control in practical applications.

9.7.3 Bypass flow-based sequence control

There is normally a set of pumps in the primary and secondary chilled water loops respectively. The two loops are hydraulically decoupled by a bypass pipe, which forms the common part of the two loops. A schematic diagram of such a piping network is shown in Figure 9.19.

   The chilled water flow rate in the secondary loop is somehow proportional to the total cooling load demand of a building. The flow rate in the primary loop, however, will only vary in steps with the number of operating chillers. The decouple bypass pipe allows any flow difference between the two loops to flow through it, thus preventing flow rate fluctuation in one loop from affecting that of the other loop.

   This strategy sequences chillers (and associated pumps in the primary loop) attempting to keep the primary system flow larger than the secondary system flow. In this way, the chilled water supply temperature in the secondary system is equal to the chilled water supply temperature in the primary system. Flow is often sensed by a flow meter in the common bypass pipe to indicate the surplus or deficit flow rate and the flow direction is sensed by a flow direction detection switch as shown in Figure 9.19. When the cooling

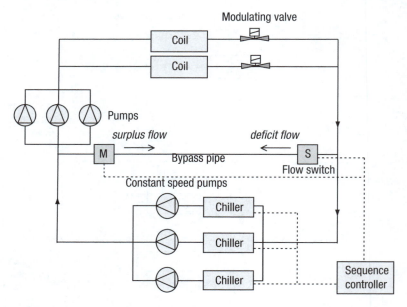

Figure 9.19 Chiller sequence control based on the flow in the bypass pipe.

load increases, the flow in the secondary system will exceed that in the primary system and cause a reverse flow direction in the common pipe, then another chiller and primary chilled water pump will be sequenced online by the sequence controller. A chiller is sequenced off when the water in the bypass pipe flows from supply to return and the flow exceeds the design flow of one chiller. Similarly, the flow direction and the value in the bypass pipe provides good reference information on building load only, and chiller sequence control based solely on the flow in the bypass pipe hardly provides precise and reliable control in practical applications.

9.7.4 Direct power-based sequence control

The simplest indicator of the simultaneous chiller cooling load is the percentage of full-load amperage of the compressor motors, not because it is such a straightforward measure of the chiller load but because it is generally available on the chiller control panel and a reliable measurement. Correlating this percentage of full-load amperage to the chiller cooling load is complicated as it relates to the influence of the power factor, part-load chiller efficiency and the capacity change due to the change of condenser and evaporator water temperatures. The percentage of full-load motor amperage tends to over-predict the actual load because the actual operation conditions may be more favourable than those of the design. For example, at 60 per cent of full-load amps the real chiller load might be only 55 per cent and at 30 per cent of full-load amps the real chiller load might be only 20 per cent. But the controller would think the instantaneous cooling load to be 60 per cent and 30 per cent, respectively. Thus the overestimation may sequence an additional chiller on earlier than it should be and hence more energy is consumed. This direct power-based sequence control recommends that the lead chiller be driven close to 100 per cent load before another chiller is brought online.

In recent years, the cost of power (kW) meters is much lower and direct power measurement is used as the regular instrumentation of many building automation systems. That means the power can be measured directly and accurately, which can be used for chiller sequence and other optimal controls more precisely. The indirect measurement (Q_{in}) of total cooling load of multiple (n) chillers can be computed using an inverse model of individual chillers as shown in Equation (9–5):

$$Q_{in} = \sum_{i=1}^{n} f_i(P_{com,i}, T_{cd,i}, T_{ev,i}) \tag{9–5}$$

where $f_i(\cdot)$ is the ith chiller inverse model, $P_{com,i}$ is the ith chiller power consumption, $T_{cd,i}$ is the ith chiller condensing temperature, and $T_{ev,i}$ is the ith chiller evaporating temperature. Having the measured total cooling load, the number of chillers required can be determined by the total cooling load-based sequence control method as described in the following section.

9.7.5 Total cooling load-based sequence control

In the above sequence control schemes, measurements of cooling load demand of a building are performed by monitoring some indicators of cooling load (i.e. the chilled water return temperature and the water flow in the bypass pipe, etc.) and therefore are indirect measurements of the cooling load. The sequence control based on the chilled water return temperature relies on the assumption that the chilled water in the secondary loop is constant and well controlled between each step, which actually is affected by the uncertainty of flow control. The sequence control based on the flow in the decoupled bypass pipe relies on the assumption that the ΔT of the chilled water to and from the building is constant. However, these assumptions are hardly maintained in practice. For example, ΔT tends to decrease below the design level for a variety of reasons, which results in more partially loaded chillers online. In addition, the above two control methods are performed according to some pre-set level of temperature or water flow rate regardless of the operating characteristics of the online chillers under different operating conditions.

In order to operate the chilled water plant in an efficient way, a direct way of measuring the cooling load is often adopted nowadays. The instantaneous cooling load can be determined through measuring the total flow rate of chilled water in the secondary loop and the difference between the chilled water supply and return temperatures in the secondary loop. The product of the two signals is then proportional to the total building load ($Q_b = \rho_w C p_w M_w \Delta T$), which should be offset by the chilled water plant.

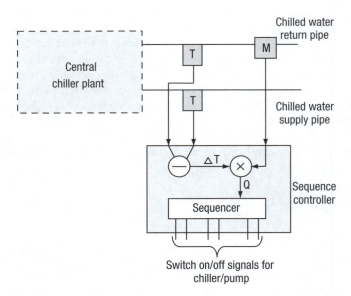

Figure 9.20 Schematics of chiller sequence control based on total cooling load.

Figure 9.20 presents a schematic diagram of the instrumentation for this control scheme.

For optimization of the chiller operation, the part-load performance of individual chillers, the combined performance of all the chillers and the chiller performance under various conditions have to be taken in consideration. Figure 9.21 illustrates the combined part-load performance of three centrifugal chillers. To optimize their operation, the number of chillers to be put into operation for a particular cooling load demand should be the one that has the lowest combined power consumption at that cooling load. To avoid the cycling on and off of chillers, the different switching points for switching on and switching off are normally used, and a minimum time for switching off (on) a chiller is set after a chiller is switched on (off).

Total cooling load-based sequence control is in principle the best approach for optimal chiller sequence control. However, the automatic controls adopting this scheme rarely operated well in practice according to surveys in Hong Kong and elsewhere. The main cause of the problem is the accuracy and reliability of total cooling load measurement using the ΔT and flow. Although there are many flow meters of high accuracy (precision of 1–3 per cent) and reasonably low cost available in the market, there is difficulty in ensuring the accuracy of flow measurements in practical installations and particularly after some period of operation due to the difficulty of site calibration and replacement. More serious errors arise from the measurement of differential temperature. The differential temperature of the chilled water loops is generally small. Its design value in most systems is around 5 K and the actual value in operation is even lower (say 3 or 4 K). An error of 0.5 K

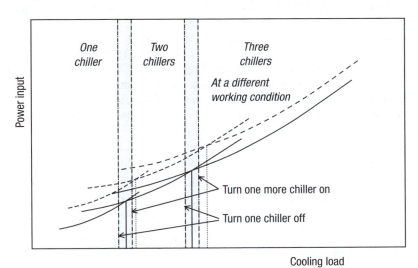

Figure 9.21 Combined part-load performance of multiple centrifugal chillers and total cooling load-based sequence control.

in temperature measurement in both supply and return water temperature sensors may result in up to 25 per cent error in total cooling load measurement if the differential temperature is 4 K. A site study was made in quite a number of 'normally' maintained chilling systems in Hong Kong by using one redundant set of sensors to measure the differential temperatures. The differences between the two total cooling load measurements using two redundant sets of sensors were over 30 per cent in a large proportion of the chilling systems investigated.

Due to the problem of using the ΔT and flow measurements, there is a tendency to use power measurement to measure the total cooling load indirectly for chiller sequence control purpose as discussed in the previous section. There are a few advantages of this arrangement. The measured variables are electrical variables, which can be measured accurately and reliably compared with measuring thermophysical variables. The measured value is an absolute value instead of a differential value. The disadvantages are obvious also. First, it is an indirect measurement. The measured chiller powers need to be transferred to the cooling load on a chiller. Normally, a reverse chiller model is needed. Second, the correlation between the power and cooling load depends on the working condition of the chiller. Therefore, the chiller condition should be included in the chiller model used. Third, this correlation may change due to chiller performance degradation.

Another way to improve the reliability of automatic chiller sequence control is to combine the above control schemes. Typically, the chilled water return temperature and the bypass flow rate (and direction) can be used as additional reference information on building cooling load and chiller operation to improve the performance of total cooling load-based chiller sequence control.

Measurement accuracy and reliability are essential for both chilling system condition monitoring and control since efficiency monitoring and chiller sequence control are particularly sensitive to them. Huang, Wang and Sun (2008) proposed a data fusion technique for the combined use of the direct cooling load measurement and indirect cooling load measurement to improve the accuracy and reliability of the cooling load measurement and, therefore, the reliability of the chiller sequence control. The method can also provide evaluation of the degree of confidence of the cooling load measurement and detect faults in measurement instruments.

9.7.6 Other issues related to chiller sequence control

For multiple chillers of equal size in a central chilling system, the load on an individual chiller should be kept the same. For multiple chillers of different sizes, the load on an individual chiller should be maintained proportional to its capacity when chillers have similar characteristics. In addition, the total operating time of individual chillers should also be kept the same, which will benefit routine maintenance and future retrofits.

The control outlined is not difficult to achieve, but it is not achieved automatically in practice. In practical operation, there needs to be certainty that different chillers have the same chilled water supply and return temperatures. That means each chiller should have the same supply temperature set-point and the water supply temperature sensors should be well calibrated (or, at least, have the same offset). The return water should be well mixed before tee-off to individual chillers. Note that a common mixing return pipe is recommended if a building has more than one return water pipe. When the same chilled water supply and return temperatures are maintained, it needs to be ensured that each chiller has the same water flow rate for chillers of the same size or the flow rate of each chiller is proportional to its capacity. This relies essentially on correct installation and commissioning.

9.8 Pump speed and sequence control of chilled water systems

Similar to chillers in central chilling systems, multiple pumps are often used in air-conditioning systems for circulation of chilled water, condenser water or sea water. Therefore, a sequence control system is definitely needed to provide control of the number of pumps put into operation to cope with the changing cooling load from time to time. For variable speed pumps in chilled water systems, a control system is needed to change the pump speed in order to provide the required chilled water circulation. Both pump speed and sequence control can save pump energy consumption. Details of the pump speed and sequence control system are determined by the characteristics of the pumps, matching between the pumps and the load side as well as the designs of chilled water distribution systems.

9.8.1 Pump sequence control through chiller and pump interlock

For primary pumps in the constant primary-only systems, constant primary/variable secondary systems and variable primary systems, the primary pump and chiller combination is designed on a one-to-one matching basis and a pump and its associated chiller are interlocked. In this case, sequence control of the pumps is accomplished by interlocking the on and off commands for a pump with its associated chiller. Lead/lag timing of pumps should be considered.

Starting of a pump that serves a chiller must take place at a time leading the starting of the chiller compressor. The goal of this lead timing is to ensure that there is already chilled water circulating in the evaporator whenever the chiller operates. Stopping time of the pump, however, must take place at a time lagging the stopping of the chiller. If the circulation of the chilled water stops as soon as the chiller compressor is stopped, the residual refrigerant in the evaporator may freeze the chilled water that stays in the evaporator tubes, thus causing them to burst. Likewise, the condenser water for water-cooled chillers must continue to flow through the condenser for a certain period of time after the chiller compressor is stopped so that the residual hot refrigerant gas discharged by the compressor will condense into liquid state, thus avoiding the build-up of excessively high pressure in the condenser.

9.8.2 Speed and sequence control of secondary pumps

In conventional constant primary/secondary variable pumping systems, control of the secondary pumps (constant or variable speed) is independent of the chiller sequence control, as the flows of two systems are decoupled. The primary concern in control of secondary pumps is to ensure that the number of pumps put into operation have adequate capacity to deliver the required amount of chilled water to the load side (i.e. cooling coils).

For sequence control of secondary pumps of constant speed, the controlled variable could be the flow rate or the differential pressure of the chilled water supplied to the load side. The signal of this variable is sensed by a flow meter installed at the main supply or return pipe of the secondary system (as shown in Figure 9.22). The differential pressure at the supply side or at a critical (or most remote) zone may be measured. When using flow rate measurement for control, the controller switches the pumps on or off according to the staged flow rate setting values pre-set. The controller decides the required number of pumps to be put into operation according to the flow rate demanded of the load side.

Figure 9.23 illustrates the control schemes of multiple constant speed pumps when the differential pressure at the supply side of the secondary chilled water loop is used as the control variable. Pressure settings for the switching points are pre-set according to the pumps' characteristic curves.

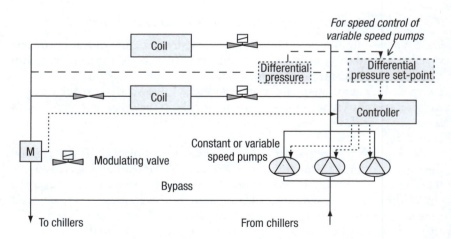

Figure 9.22 Schematic diagram of sequence control for secondary pumps in constant primary/secondary variable pumping systems.

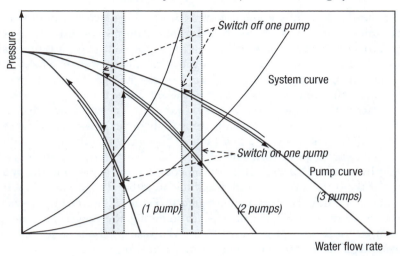

Figure 9.23 Settings of sequence control of constant-speed secondary chilled water pumps based on differential pressure.

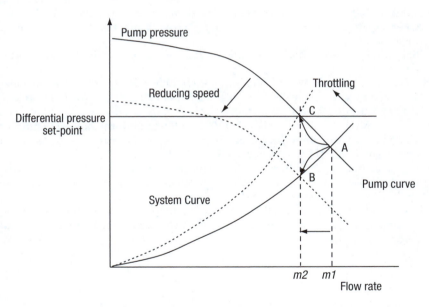

Figure 9.24 Control of variable-speed pump with pre-set differential pressure across the critical load branch.

In the strategy illustrated in Figure 9.23, a lag is specified to avoid the cycling on/off of pumps and the flow rates at switching on points are slightly higher that that of relevant switching off points.

For speed control of secondary pumps of variable speed, the pumps act as distribution pumps and are controlled to maintain pre-set differential pressure across the critical load branch or at the supply side while it is preferable to use the differential pressure across the critical load branch. The optimal set-point of differential pressure should be the least differential pressure allowing all the cooling coils to have sufficient water flow. When the measured differential pressure exceeds the set-point, the speeds of variable speed pumps will be reduced by the controller. When the measured differential pressure decreases below the low limit of the set-point, the speeds will be increased. Note that the flow rate of the chilled water loop is primarily controlled by modulating the valves in the cooling coils (changing the system curve) if the differential pressure is sufficient, although the pressure setting will affect the flow control as illustrated in Figure 9.24. The optimization of the pressure set-point is primarily to minimize the power consumption of pumps. For this purpose, the differential pressure should be set as low as allowed while being still sufficient for the critical branches.

References

ASHRAE. (2007) *ASHRAE Handbook – HVAC applications* (Chapter 41), Atlanta, Georgia: American Society of Heating, Refrigerating and Air-Conditioning Engineers, Inc.

CIBSE. (2000) *CIBSE Guide H – building control systems*. Oxford: Butterworth-Heinemann.

Huang, G. S., Wang, S. W. and Sun, Y. J. (2008) 'Enhancing the reliability of chiller sequencing control using fused measurement of building cooling load', *HVAC&R Research*, 14(6): 941–58.

Ma, Z. J., Wang, S. W. and Xiao, F. (2009) 'Online performance evaluation of alternative control strategies for building cooling water systems prior to in-situ implementation', *Applied Energy*, 86(5): 712–21.

Sun, J. and Reddy, A. (2005) 'Optimal control of building HVAC&R systems using complete simulation-based sequential quadratic programming (CSB-SQP)', *Building and Environment*, 40(5): 657–69.

Wang, S. W. and Burnett, J. (2001) 'Online adaptive control for optimizing variable-speed pumps of indirect water-cooled chilling systems', *Applied Thermal Engineering*, 21(11): 1083–1103.

10 Lighting-control systems

Artificial lighting is essential for the visual environment in spaces for living, working or other generic purposes where and when there is no sufficient daylight available. In some special spaces, such as spaces for entertainment, lighting is needed for creating a dramatic or dynamic environment. The lighting system is one of the major energy consumers in buildings, typically following the HVAC system in office and commercial buildings. The energy efficiency of lighting systems and quality of the visual environment provided are determined by the selection of lamps (including the associated components), the architectural layout and the control. This chapter focuses on the control systems of lighting for generic purposes.

10.1 Purpose of lighting-control systems

The external daylight condition, the occupancy and the use of a space are changing. Control of the lighting system is required to meet the following purposes typically, which may be achieved manually or automatically:

- functional need and flexibility of the space;
- energy saving;
- visual comfort of the occupants;
- the requirements of legislation;
- creating a dynamic or dramatic environment.

Different requirements of the visual environment are needed for different activities or functions. For instance, a lecture theatre is often used for PowerPoint presentations, so the luminance level in the space should be at a lower level, allowing the audience to see the slides presented on the screen clearly and comfortably. When the lecturer is speaking without a PowerPoint presentation, a higher luminance level is preferred to allow pleasant and effective communication. If an audience is entering a movie theatre and people are finding their seats, the lights should be on, but they should be off in order to allow the audience to see the film properly when the movie starts. It is often important to provide different lighting in a space to allow for different functional uses. The lighting system should also be adaptable to the changes in the space, such as office layout.

Energy efficiency is one of the important issues concerning lighting system control. A very significant proportion of building energy is consumed by the lighting systems. Providing lighting only in the areas and in the periods lighting is needed and providing the right level of lighting as needed are effective means of reducing the energy use of the lighting system. The main control actions for this purpose are on/off switching and dimming.

Lighting is one of the major contributors to create a stimulating and comfortable environment for working and living. It is one of the major environmental factors affecting the satisfaction of the occupants in residential buildings. Lighting is also an environmental factor affecting the productivity of the occupants. Different people also prefer different levels of lighting. The visual environment adaptable to individual requirements or controllable by individuals will also increase the satisfaction of users with the visual environment and lighting systems.

Lighting has become a subject of legislation in many countries. Typical concerns are: setting standards for certain types of spaces and activities to ensure its effectiveness; setting standards to ensure personal safety and security; setting standards on the use of lighting technology and systems to ensure energy efficiency.

Creating the desired aesthetic effects of spaces in practical fashion has been the major driving force for the lighting control market, although energy management is playing an increasing role. Lighting systems can be controlled to provide a balance of different light sources to ensure a pleasing visual environment or the transition between one light state and another. Depending on the functional purpose of a space, lighting systems may be controlled to provide 'comfort' in the sense of a relaxing or pleasant visual environment. Lighting systems may also be controlled to provide a dramatic and dynamic environment.

10.2 Basic components of lighting and lighting-control systems

10.2.1 Lamps

Typical lamps used for generic purposes include the incandescent lamp, tungsten halogen lamp, fluorescent lamp and compact fluorescent lamp. Other types of lamps include lamps for high-intensity discharge (HID), such as high pressure mercury lamps, high pressure sodium lamps, metal halide lamps, xenon lamps and lamps for special purposes.

The invention of the incandescent lamp in the late 1870s started the practical use of artificial electric lighting. It was the main electric light source until the recent application of more competitive light sources. Even today, it is still very widely used, particularly in home applications.

The fluorescent lamp is one of the large family of discharge lamps. In these lamps, light is generated by an electrical discharge within a gas or vapour. Straight tubular lamps have been the main type of fluorescent lamps and are still among the most commonly used fluorescent lamps today. Long life and high efficiency are the main features of fluorescent lamps. Recent developments in the electronics industry allow electronic components (typically ballasts) to be made very small in size to serve as the internal components of a fluorescent lamp. This has allowed the lamps and associated electronic components to be integrated as compact fluorescent lamps. Such lamps are nowadays very widely used in the traditional applications of incandescent lamps and special lamps.

10.2.2 Ballasts and dimming ballasts

Ballasts are the devices required to control the starting and operating voltages of electrical gas discharge lamps, such as fluorescent lamps, neon lamps and HID lamps. The lighting ballast is used to limit the flow of current through a lamp, which can be a very simple resistor or rather more complex devices (such as electronic ballasts).

Ballasts are necessary to operate discharge lamps because they have negative resistance, meaning they are unable to regulate the amount of current that passes through them. Therefore a ballast must be used to control current flow, otherwise the lamp could fail. Electromagnetic ballasts use electromagnetic induction to provide the starting and operating voltages of gas discharge lamps.

Electromagnetic ballasts limit the flow of current to the light but do not change the frequency of the input power. The lamp then illuminates on each half-cycle of the power supply. This is why many fluorescent and neon lights flicker visibly. Since the light illuminates on half-cycles, the rate of flicker is twice the frequency of the power source, with a result that the light flickers at 100 Hz or 120 Hz. A more modern type of ballast is electronic instead of electromagnetic. Electronic ballasts use solid state circuitry to transform voltage. But unlike electromagnetic ballasts, they can also alter the frequency of power. This means that electronic ballasts can greatly reduce or eliminate any flicker in the lamps. Because it uses solid-state circuitry instead of magnetic coils, it is also more efficient and therefore runs at a lower temperature.

The ballast can also be configured to change the current flowing through the lamp while receiving a signal from a control device and afterwards, and consequently achieving a gradual controlled reduction in lamp output. Dimming ballasts are available for both linear and compact fluorescent lamps. Dimming fluorescent lamps can provide significant benefits to owners of commercial lighting systems including the following:

- *Flexibility:* allowing the lighting system to adapt to multiple activities and the change of space use.

- *Energy saving:* resulting from direct energy savings as well as load reduction during peak demand periods.
- *Higher comfort of occupants:* satisfaction and comfort achieved by allowing occupants to choose their own light levels or controlling the light level at a certain desired range.
- *Increased lamp life:* for applications where lamps can be dimmed instead of frequently switched on or off.

The available control range of dimming ballasts is typically 25–100 per cent, 10–100 per cent or 5–100 per cent. Whichever is chosen should be sufficient for energy saving as well as architectural lighting purposes. Some applications might demand ballasts with a wider dimming range such as 3–100 per cent or even 1–100 per cent.

The dimming ballasts can be categorized into two types: analogue electronic dimming ballast and digital electronic dimming ballast. The most popularly used analogue method is 0–10 V DC as the control input to the dimming ballast. The digital electronic dimming ballast includes a more functional component, a microprocessor. It performs as a storage, receiver and sender of digital information. The microprocessor stores the ballast address, receives control signals and sends status information.

Both analogue and digital dimming ballasts provide the essential function of controlling the lamp output based on input from a control device. Both enable the construction of networks of controls and ballasts wired to local and central points where control signals can originate, either manually or automatically.

Analogue dimming ballasts have been in the market longer, typically presenting a lower cost and compatibility with a wide range of common control devices. Digital dimming ballasts provide a higher degree of control capability, such as the ability to individually address and group the ballasts, gain feedback information from ballasts and provide the flexibility of lighting systems in responding to changes of use over time.

10.2.3 Dimmers

Dimmers are devices used to vary the brightness of lamps. Dimmers vary the intensity of the light outputs of lamps by decreasing or increasing the voltage and therefore the power to the lamps. Dimmable ballasts of a fluorescent lamp serve the same purpose of varying the light outputs but they are usually classified as their own standalone category, dimmable ballast.

Typical examples of early non-electronic dimmers are resistance dimmers, reactor dimmers and transformer dimmers. Typical examples of modern dimmers are dimmers based on thyristors, transistors or silicon-controlled rectifiers (SCR). Dimmers range in size and capacity from the size of an ordinary light switch used for residential lighting to high power and multiple-channel units used in architectural or theatre lighting applications. Modern professional dimmers are usually controlled by a digital control system via control networks.

Non-domestic dimmers are usually controlled remotely by means of analogue or digital control. Analogue-controlled dimmers usually require a separate wire for each dimming channel to receive a control signal of a voltage between 0 and 10 V. Modern dimmers have built-in microprocessors to receive the control signals from controllers and convert the digital signal directly into a control signal for the switches while providing the opportunity for diagnostic feedback to be sent digitally to the lighting controller. The analogue and digital protocols will be discussed late in this chapter.

10.2.4 Sensors and control devices

Light sensors are a basic element for the automatic control of modern lighting systems. A light sensor measures the level of light in a room or ambient light for dimming or switching the lights. There are many types of light-sensitive materials used for light level measurement. The most commonly used light sensors used nowadays for light control applications are based on a silicon photo-diode. The silicon photo-diode is typically used in a photovoltaic mode which is coupled to an amplifier to provide either a proportional (analogue) output or a binary digital signal for operating a relay. Typical light sensors used for light control application offer a measurement range between a few dozen lux and a few thousand lux (such as 20–3,000 lux).

Light sensors may be designed simply to provide the analogue output (i.e. 0–10 V) or binary digital output for the use of a controller. Light sensors might be designed to provide some simple control of lamps such as to provide a 1–10 V analogue output to directly control electronic ballast, or to provide an on/off balance output for the on/off control of a switch. A communication interface may be also coupled to the light sensor allowing the light measurement output to be sent to controllers or the control signal to the control unit via the control network.

An electronic *motion detector* contains a motion sensor that transforms the detection of motion into an electric signal. This can be achieved by measuring optical or acoustical changes in the field of view. A motion detector may be connected to a burglar alarm that is used to alert the home owner or security service after it detects motion. A motion detector can also be used for lighting control, such as an *occupancy sensor*. It senses when motion has stopped for a specified time period in order to trigger a light-extinguishing signal. These devices prevent the waste of lighting energy in unoccupied spaces.

There are basically three types of sensors used in motion detectors: spectrum, i.e. passive infrared sensors (PIR); ultrasonic (active); and microwave (active). The PIR sensor looks for body heat and no energy is emitted from the sensor. The ultrasonic sensor sends out pulses and measures the reflection off a moving object. The microwave sensor sends out microwave pulses and measures the reflection off a moving object.

Dual-technology motion detectors are used for applications such as security systems, which employ a combination of different sensing technologies. These dual-technology detectors benefit from each type of sensor, and false alarms are reduced. PIR technology is often paired with another model to maximize the accuracy and reduce energy use. PIR draws less energy than microwave detection, and so many sensors are linked so that when the PIR sensor is tripped, it activates a microwave sensor.

10.2.5 Analogue control and digital control

In the past, light dimmers and switches were all operated manually. A leveller or knob was used to interrupt the supply directly or to move a sliding contact. Analogue control and digital control are the two basic means for the remote or automatic control of lighting systems today.

Analogue control is obviously the simplest method of remote or automatic control of dimmers and other control actuation devices that vary the voltage or other parameters of the power supply to lamps. There are many types of signals used for lighting control similar to the control of other systems. Nowadays, two standard ranges of analogue signals are usually used for lighting-control systems: 0–10 V and 1–10 V.

The *0–10 V Analog Control Protocol* is defined by the Entertainment Services and Technology Association (ESTA) in the United States. Practically all the entertainment dimming devices offering analogue control follow the specifications proposed by this protocol. The protocol specifications are listed as Table 10.1 for both controlled and controlling devices respectively.

The *1–10 V* standard used for the control of ballasts for fluorescent lamps is an extension of analogue control standard IEC 60929. It specifies the control range of 1–10 V to ensure that electrical noise does not affect system control performance. The standard specifies only the 'minimum level',

Table 10.1 Specifications of 0–10 V analogue control of ESTA

Controlled devices		Controlling devices	
Control range	0–10 V	Passive control source impedance	<10 kΩ
OFF control voltage	≤0 V	Active control source impedance	<10 kΩ
100% or ON control voltage	≥10 V	Current source capacity	>2 mA
Control	Linear	Stability at constant output	±20 mV
Safe acceptance range	−0.5 to +15 V	Diode blocking capacity	≥10 V
Input impedance	100 kΩ ±20%		

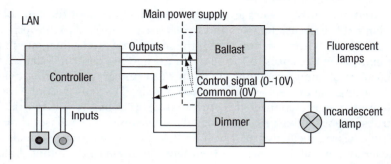

Figure 10.1 An example of conceptual configuration of analogue lighting control.

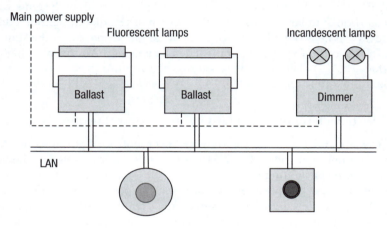

Figure 10.2 An example of conceptual configuration of digital lighting control.

but does not specify any 'OFF' command. Separate provision is needed for 'OFF' control and for load circuit isolation. It is worth noting that 0–10 V is also widely used as the analogue control signal for electronic ballasts of fluorescent lamps.

Prior to the implementation of LAN-based digital systems, some systems used a multiplexed analogue control in order to use 'one wire' to control multiple dimmers. However, this type of technology is being phased out today and digital control technology using a network concept is widely used. The aim of controlling many lights through one wire is now achieved using more effective network-based digital control technology.

Nowadays, analogue control is still used widely in lighting control for sensor signals and control signals. For large-scale lighting-control systems, digital control should be used. In fact, digital control is also finding increasingly wide application in sensor signals and control signals. Figure 10.1 illustrates an example of conceptual configuration of a lighting-control system using a standalone controller employing analogue control. The sensors and external control devices are connected to the inputs of the controllers. The dimmable ballasts, dimmers and switches are connected to the outputs of the controllers. It is worth noting that the controllers can be integrated using communication networks to form large-scale lighting-control systems.

Figure 10.2 shows the configuration of an integrated lighting-control system using digital sensor and control signals. The sensors and external control devices are connected to the control stations via a network. The control stations send their control signals to the dimmable ballasts, dimmers and switches via the network as well. In this system, the control signals may be generated by the smart sensors themselves and sent to the remote ballasts or dimmers via the network.

There have been many proprietary protocols used or developed for lighting-control systems. Fortunately, most of the lighting-control systems today are based on standard protocols. In today's market, digital lighting control products can be grouped into two categories. One is based on the standard protocols specific for lighting control, such as Digital Addressable Lighting Interface (DALI). The other is based on the standard protocols generic for automation applications.

10.3 Systems based on standard lighting-control protocols

DMX512-A and *DALI* are the standard protocols for lighting-control systems in popular use today. DMX512-A was originally targeted at the needs of entertainment lighting control while it offers high channel capacity and high communication speed and requires some experience in its implementation. DALI is targeted at the needs of simpler commercial and architectural lighting applications and requires less experience in its implementation. *MIDI (Musical Instrument Digital Interface)* is a widely used protocol in the entertainment industry.

10.3.1 DMX512-A

DMX512-A communications protocol is based on EIA-RS-485 (RS-485), which is most commonly used for stage lighting and effect lighting controls. It was originally developed by the United States Institute for Theatre Technology (USITT) in 1986 and revised in 1990 as USITT DMX512/1990. The Entertainment Services and Technology Association (ESTA) started to revise the protocol in 1998 and developed the standard as an ANSI standard known as Entertainment Technology – USITT DMX512-A, approved by ANSI in November 2004. It is usually called simply DMX512-A, and maintained by ESTA.

The DMX512-A connection is based on a connector with five pins (namely five-pin XLR), though only three pins of the five are always used. The first pin is for the signal common. The second and third pins provide the primary data link, the second pin as the drive complement (–) and third pin as the drive true (+) respectively. The fourth and fifth pins provide the secondary data link, which is optional, similarly as the drive complement (−) and the dive true (+) respectively.

The cabling specification was removed from DMX512-A after 2004, although it is essential. It is now specified by separate standards. DMX512-A uses the concept that data is sent in packets. Each DMX512-A data link transmits a start code that identifies the data type and up to 512 8-bit values, between 0 and 255, so one cable typically controls 512 dimmers or devices. In cases where more than 512 control channels are needed, multiple DMX universes can be used.

DMX512-A data is sent using EIA-485 voltage levels and cabling practices. Data is transmitted serially at 250 Kbps and grouped into packets of up to (but not necessarily) 513 bytes each. A full packet takes approximately 23 ms to send. That corresponds to a refresh rate of about 44 Hz if the maximum number of 512 channels is used. If a higher refresh rate is needed, fewer channels should be used.

The DMX512-A uses 8 bits (256 levels) for each channel. But it does not specify the correlation between the lighting level and the data (bits). Any correlation relating the data and lighting levels is carried in the dimmers. DMX512-A's popularity is partly due to its robustness. The cable can be abused without any loss of function in ways that would render Ethernet or other high-speed data cables useless.

It is worth noticing that RS-485 is used as the Physical Layer underlying many standard and proprietary automation protocols used to implement industrial control systems, such as Modbus and PROFIBUS, as well as DXM512-A. RS-485 only specifies electrical characteristics of the driver and the receiver. It does not specify or recommend any data protocol. RS-485 enables the configuration of inexpensive local networks and multidrop communications links. It offers high data transmission speeds (35 Mbps up to 10 m and 100 Kbps at 1200 m). Since it uses a differential balanced line over twisted-pair cable (like EIA-422), it can span relatively large distances (up to 1200 m).

10.3.2 DALI – Digital Addressable Lighting Interface

It has been an aim, for which effort has been expended, that within a lighting system every light is separately controllable but only a single control cable is sufficient for all devices in the system. The LAN technology discussed in Chapter 5, such as LonTalk, EBI and BACnet, actually made it possible to achieve such an aim. One problem is that such a system may have a high cost per node. DALI has therefore set its ambitions low. It is not targeted to be suitable for the control of various building services systems. Instead, it seeks to be the optimum method (in terms of cost and capacity) for controlling lights within a large space or a number of rooms in commercial, architectural lighting applications.

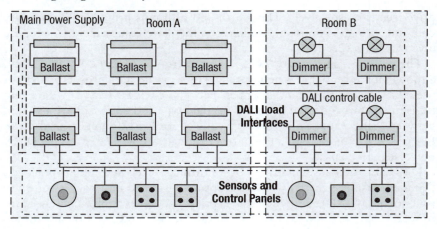

Figure 10.3 An example of lighting-control systems using DALI protocol.

Work associated with the DALI project began in the mid-1990s. The original intention, as agreed by the European ballast manufactures, was to ensure standard control of digital ballasts. A few manufacturers introduced their concepts in 1999. Since then, many companies have followed suit. In the last few years, DALI has been a popular technology in the lighting-control systems market.

The conceptual architecture of a DALI system is shown in Figure 10.3. Each lamp is associated with a digital load interface, such as an electrical ballast. The control devices (e.g. push-button panels) and the sensing devices (e.g. occupancy detectors) are linked to their associated load interfaces or controllers in the system via a single control cable (a simple pair of wires). It is possible for an individual control device to control one specific lamp or a group of specific lamps. It provides the flexibility to re-couple the control and sensing devices with the lamps in case partitions of an office are reorganized.

Table 10.2 summarizes the basic characteristic data of DALI. DALI technology provides a very low data transmission rate that restricts its application in commercial, architectural and residential applications. The advantage, as a result of the low transmission rate, is that there is no need for a rule on the wiring of the control cables in terms of topology and line termination. DALI offers a wide range of signal tolerance, resulting in the robustness of the system to electrical noise. The allowable ranges of LOW signal are –4.5 V to +4.5 V and –6.5 V to +6.5 V for transmitting and receiving respectively. The allowable ranges of HIGH signal are +11.5 V to +20.5 V and +9.5 V to +22.5 V for transmitting and receiving respectively. Complete isolation of the control lines from the main power supply is provided.

The DALI protocol is specified to allow an 'OFF' signal so that the control device can turn off a light completely. The address bits of the devices

Table 10.2 Basic characteristic data of DALI

Maximum number of nodes in a single network	64
Transmission speed	1200 bps
Data coding method	Manchester
LOW signal	0 V (nominal)
HIGH signal	16 V (nominal)
Maximum control cable length	300 m
Signal supply current limit	250 mA
Nominal signal per node	2 mA
Number of levels	255 plus OFF (8 bit)

are arranged to provide the flexibility of re-coupling the controlling devices and the load interfaces and changing the grouping of load interfaces without changing the address number of ballasts in terms of controlling and pulling for backward data. The ability to extract status data from the load interfaces is an important feature of DALI, which is done on the basis of pulling. Important status data of a load, such as 'lamp failure', on/off status, light level and load current, can be collected. This data is very useful for a BAS in order to set up system monitoring and diagnosis to enhance the lighting system performance.

Due to the popular use of DALI technology in recent years, many DALI components are available from different manufacturers. Typical examples of available components are: load interfaces (e.g. dimmers, electronic transformers, electronic ballasts and blind and curtain controllers), control panels (on/off control, dimming control, etc. for manual control), sensors (e.g. light sensor, occupancy sensor, etc. for automatic control), and control interfaces.

A control interface provides the flexibility and ability of a DALI system to use non-DALI lighting components or integrate with other building automation systems. An example of DALI components in this category include 1–10 V analogue converter (allowing analogue controlled dimmable ballast to be used in lighting systems using DALI control systems), DALI/LonWork converter and DALI/EIB converter.

10.4 Systems based on common automation protocols

Chapter 5 introduced a numbers of standard protocols popularly used in building automation systems. Some of these protocols, including other automation protocols, are used in the lighting control industry as protocols at field network level. Some are used for integration of the lighting-control systems or integration of the lighting-control system with other BA systems.

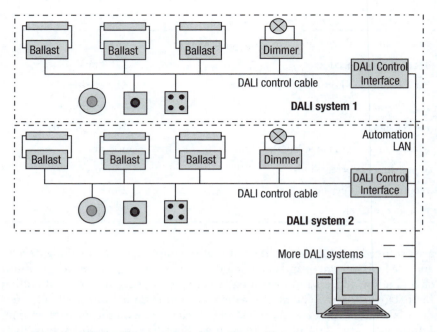

Figure 10.4 Example of large DALI lighting-control system integrated using automation LAN.

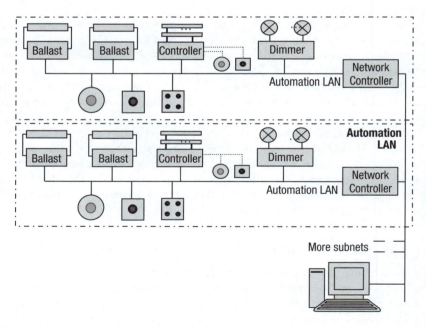

Figure 10.5 Example of large lighting-control system based on automation LAN.

Commonly used automation protocols include LonWorks, EIB and C-Bus. A summary of LonWorks and EIB was given in Chapter 5. A brief outline of C-Bus is given here. *C-Bus* is used in the control of home automation systems and commercial building lighting-control systems. It uses a dedicated low-voltage cable (unshielded twisted pair) or two-way wireless network to carry command and control signals. This improves the reliability of command transmission and makes C-Bus far more suitable for large commercial applications. The transmission rate of networks using C-Bus may range from 4.8 to 920 Kbps. A typical example of a C-Bus network for lighting application is 9.6 Kbps. The C-Bus network wiring uses 'tree topology' architecture. The maximum length of cable used on a C-Bus network is 1,000 m, and up to 100 units can be installed on a single C-Bus network. The maximum number of C-Bus networks in one installation is 255. C-Bus has attracted applications mainly in Australia but also in Europe, Asia and the United States. There is a similar protocol called CEBus for similar use but with more North American applications. The main difference is that CEBus provides a power-line carrier option.

The conceptual architectures of the lighting-control systems using generic automation protocols (e.g. LonWorks or EIB) are similar to that of the systems using DALI. The main differences between them are as follows. A DALI node is associated with one channel load (load, control or sensor, etc.) while one node in the lighting-control systems using those generic automation protocols may be designed to handle multiple channels due to the network capacity and characteristics. The speed of the DALI system is very low, whereas most of the automation protocols have a higher speed. The maximum number of units that a DALI system can handle is limited. To integrate more of a DALI system to form a large lighting-control system needs other network technology, such as LonWorks, BACnet, EIB or Ethernet/IP, at the higher level supported by DALI converters.

Figure 10.4 shows an example of a large DALI lighting-control system integrated using automation LAN. It can be seen that all the load interfaces, sensors and control devices are smart devices which are connected to the DALI control line directly. Many different DALI devices can be selected depending on availability from manufacturers. However, one important feature of the DALI system is that each of the load interfaces, sensors and control devices is an independent node in the DALI system. If a large system needs to be formed, more DALI systems should be integrated via other automation networks. The centralized computer monitoring and management can be achieved via such an automation network while computer monitoring of an individual DALI system can also be made directly using DALI interface, such as a serial converter for connection to a computer.

Figure 10.5 shows an example of a large lighting-control system based on automation LAN. It can be seen that a load interface can be either a single-channel device or a multiple-channel device. This means that one controller can control a number of ballasts. A single-channel smart ballast may be connected to the automation LAN directly. A dimmer may handle a number of light loads. A sensor or other control device may be a smart device which can be connected to the automation LAN directly. It may be linked to a controller. Many different devices can be selected and many different system configurations can be designed depending on the availability of devices from manufacturers. If a large system needs to be formed, more subsystems should be integrated via the same automation LAN or other (higher speed) automation LAN. The integration of the lighting-control system could be carried out with other BA systems at the subnet level directly or at the higher level depending on the compatibility of the networks and the need for interoperation.

10.5 Strategies for energy management and lighting control

The control objective of lighting systems in spaces for living, working or other generic purposes is to provide the optimal visual environment for the visual comfort and productivity of occupants with a minimum of energy consumption. The typical approaches to achieve such an objective may be summarized as follows:

• providing lighting where lighting is needed;
• providing lighting when lighting is needed;
• providing the right amount of lighting;
• making use of daylight as much as possible.

The most effective means of saving energy in buildings is to turn off the systems where and when they are not needed. The run-up time for many general-purpose lamps is either negligible (such as incandescent lamp) or very short (such as fluorescent lamps). That makes it practical to turn off the lighting systems in a space when it is not occupied as it can provide the lighting services almost immediately after it is switched on when the space is occupied.

To control the lighting system providing the service only where lighting is needed, proper zoning or grouping of lighting fixtures is essential. The zoning of lighting fixtures is typically established according to the functions of the spaces, lighting schedules, architectural layout, availability of daylight, and so on. There is a need to compromise on sizing the control zones of the lighting systems correctly due to the fact that smaller control zones generally result in greater control resolution, greater energy-saving potential as well as greater opportunity to enable the lighting system to meet the lighting needs. But smaller control zones generally will make the lighting-control system more complicated in terms of operation and control, and more expensive in terms of initial installation and maintenance.

Providing light only when it is needed can be achieved typically by including scheduled control and occupancy-based control. For instance, if the use of the spaces is predictable according to a space-booking system, a time-based strategy can be considered and the lighting of the spaces can be scheduled automatically. Monitoring the occupancy of the spaces is another typical means for lighting control nowadays. Typical occupancy-monitoring techniques are the use of occupancy detectors. If access control of the spaces is available and integrated with the lighting-control system, the lighting-control system can obtain the information on occupation of the spaces from the access-control system.

Providing the right amount of lighting may be achieved simply by proper switching of lights in the zones. But dimming control is a much more precise means of controlling a lighting system to provide the right amount of lighting for the functional need of the space. The maximum use of daylight reduces the need for artificial lighting and therefore energy consumption. Shading control as a special function of the lighting-control system enables the effective use of daylight and provides visual comfort. A few basic technologies and options for energy-effective lighting control are discussed in the following sections.

10.5.1 Switching and dimming controls

Switching and dimming controls are the two basic means of controlling light outputs for the purposes of energy management and visual needs. The main benefits of dimming control are better occupant satisfaction and comfort and flexibility. The main advantage of switching control is that it is relatively inexpensive and simpler for commissioning. The main advantages of dimming control is that the light output can be set at any level and greater user acceptance can be achieved due to smooth transitions between light levels. The main disadvantage of switching control is lower user acceptance in occupied spaces with stationary tasks due to abrupt and noticeable changes in light level. The main disadvantages of dimming control are higher installation costs and more complications in commissioning.

10.5.2 Scheduled and sensor-based controls

If the occupation of spaces is fixed or predictable, the lighting-control system can be programmed to switch the lighting on or off in spaces, including the lighting levels, according to a schedule. These schedules can be fixed, which may be different for different days in the week or on working days and holidays. The operators may update or set these schedules according to the actual booking of the spaces. In the case where the lighting-control system is integrated with the space-booking system or access-control system, these schedules can be set and updated automatically according to the actual booking or use of the spaces.

Nowadays, sensor-based control is very widely used to control lighting systems automatically, which is based on monitoring the occupancy in the spaces. Typical occupancy-monitoring techniques are the use of occupancy sensors. Occupancy sensors usually employ motion detectors. A summary of the types of occupancy sensors is given in Section 10.2.4.

Sensor-based control is also essentially used for both switching and dimming controls to reduce the energy consumption of a lighting system by maximizing the use of daylight. Open-loop control is typically implemented by sensing the ambient daylight level. Closed-loop dimming control is also commonly used by sensing the actual light (lux) level of the served indoor spaces.

10.5.3 Use of daylight and blind/shade control

Making the best use of daylight is an obvious means of saving energy in lighting systems. Glass is a key element in the architectural expression of the building and typically provides indoor spaces with a visual connection with the outdoors and daylight to enhance the quality of the indoor environment. But the building skin must serve a crucial function in its role to help maintain proper interior working environments under extremes of external environmental conditions. Solar and daylight fluxes can vary very rapidly over a very wide range. Controlling solar gain and managing daylight, view and glare are issues for the effective use of daylight, and are still a challenge today.

Traditional manually operated mechanical shading systems such as blinds or shades can be motorized and then controlled by occupant action or by sensors and building controls. Emerging smart glass technology can even dynamically change optical properties, and can be activated manually or by automated control systems. In these cases, electric lighting should be controlled as the backup to meet occupant needs, while maximizing energy efficiency and minimizing electricity demand.

Manual operation of windows or shades may work well in homes and some small buildings. But in a larger building with many occupants and which aims to integrate the façade and lighting systems as well as HVAC systems, more reliable automated controls are essential.

Motorized blind systems, such as venetian blind systems, are well-established technologies for controlling solar gain and glare. Because both the optical properties of the slats and their tilt can be controlled there is a wide range of optical control available. Smart control on the automated blind systems can keep direct sun out of the space, reducing glare and cooling loads. Motorized shade systems, such as roller shade systems, can provide a wide range of solar optical properties. Although mechanically simpler, the shade systems have more limited optical control than blinds in terms of position, although it is also possible to layer blinds or use various fabrics.

References

DiLouie, C. (2006) 'Introduction to lighting control'. Available at: www.aboutlightingcontrols.org/education/papers/introduction.shtml

Selkowitz, S. and Lee, E. (2004) 'Integrating automated shading and smart glazing with daylight controls', International Symposium on Daylighting Buildings (IEA SHC Task 31), Tokyo, Japan.

Simpson, R. S. (2003) *Lighting Control: technology and applications*, Oxford: Focal Press.

Wikipedia. (2009) 'DMX512-A'. Available at: http://en.wikipedia.org/wiki/DMX512-A

11 Security and safety control systems

Security and safety consist of those measures taken by an organization to provide the protection of property, life, materials and facilities against fire, damage, unauthorized entry, theft and any other dishonest, illegal or criminal acts that might happen to the organization. Architects and engineers aim to find cost-effective solutions addressing the security and safety concerns of employers, employees, customers and other users of buildings, which not only comply with government regulations, but also provide enhanced protection. This chapter provides an introduction to the systems and their main technologies, including:

- closed circuit television (CCTV) systems;
- access control systems;
- burglar alarm systems;
- fire alarm systems.

11.1 CCTV systems

For decades, CCTV has been implemented and integrated in safety and security applications. The purpose of CCTV in security solutions is to provide remote 'eyes' for security operators by providing live-action displays from a distance and/or to keep a video record of the spaces under monitoring. With today's labour costs, CCTV is a cost-effective means for expanding security and safety control. Certainly, the main objective of CCTV systems should not be to record 'thieves', but rather to prevent theft. There are two basic categories of CCTV systems: *analogue CCTV systems* and *digital CCTV systems (or IP surveillance systems)*.

11.1.1 Basic CCTV components and analogue CCTV systems

A typical analogue CCTV system includes the basic components of camera, monitor, video switcher and video recorder. One physical device may integrate the functions of two or more components among monitor, video switcher and video recorder.

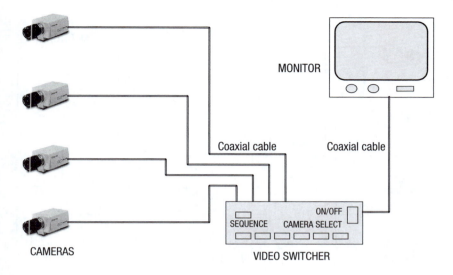

Figure 11.1 A line-powered CCTV system without recording.

A *camera* is the basis of any CCTV system. The camera creates the picture that will be transmitted to the monitoring or control position. However, CCTV is not just a simple camera-cable-monitor arrangement. The basic function of a CCTV camera is to convert the physical scene viewed by the camera into an optical picture. By a focusing process, the scene is placed upon a special camera imaging tube which scans the imaged scene and breaks it down into various picture elements. These elements are then transmitted and converted into varying illumination levels that correspond to the video signal, which is ultimately converted into a visual scene on the system monitor. The CCTV system should have the means of recording, such as using VCR (video cassette recorder), computer disc or other storage media, to maintain permanent records of what the cameras have seen. The recorded information could be used for future investigations or as evidence in prosecutions.

Cameras used in CCTV systems may be installed with lenses of fixed length. Many cameras used in CCTV systems are movable with lenses of adjustable length as in many applications the areas to be covered would need many fixed cameras. The cameras can be controlled to rotate horizontally and vertically from a remote location. In conventional analogue CCTV systems, the cameras are linked to the video monitor or video switcher via coaxial cables and each camera needs its own cable connection.

The camera may obtain power from the video switcher or monitor for its operation via the same coaxial cable for video signals, known as the line-power camera. The camera may alternatively get power by connecting to the main power supply, known as a mains-powered camera. Figure 11.1 illustrates the basic configuration of a small-scale line-powered CCVT system

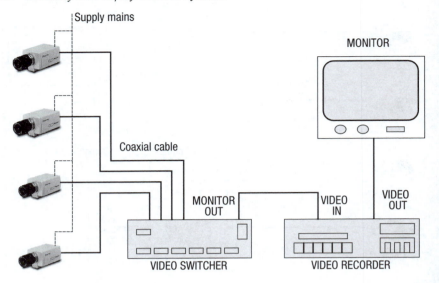

Figure 11.2 A mains-powered CCTV system with recording.

without recording. Figure 11.2 illustrates the basic configuration of a small-scale mains-powered CCVT system with recording. The arrangement of a mains-powered CCTV system allows much more flexibility in designing complex CCTV systems.

The picture created by the camera needs to be reproduced at the control position. A CCTV monitor is virtually the same as a television receiver (a TV could actually be used if it is cost effective) except that it does not need or have the tuning circuits. When multiple cameras are involved in a CCTV system, video switchers are required as shown in Figures 11.1 and 11.2. Using a video switcher, the picture of any camera may be selected to be displayed on the screen or it can be set to display the pictures of the cameras in turn according to pre-set speed and sequence. The video switcher may allow the display of the pictures of multiple cameras on the screen at the same time.

11.1.2 IP surveillance systems

Due to the popular use of broadband IP networks in buildings and digital video recording in recent years, digital surveillance systems using an IP network are rapidly becoming popular in security applications as they provide a bridge to enter the digital world with solutions of high performance, low cost and high flexibility. For the past two or three decades, video monitoring and surveillance applications have been dominated by analogue technology, and video has traditionally been recorded to VCRs.

Digital surveillance systems offer many obvious advantages over the analogue systems, such as remote accessibility, ease of use and expending, search

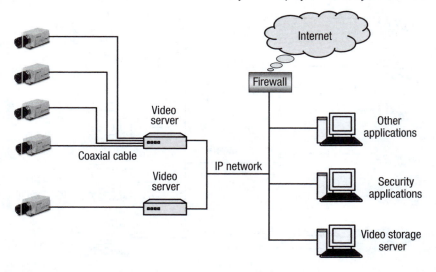

Figure 11.3 An IP surveillance system using analogue cameras supported by video servers.

capabilities, simultaneous record and playback, no image degradation, large storage capacity, integration potential and the possibility of using the existing network and sharing the network with other BA systems.

The main benefits of digital surveillance systems derive from their digital nature. This means many functions, which are impossible or difficult to provide with analogue systems, can be processed conveniently by the computing facilities. Flexible and proactive image distribution is one example of such functions. Another example is automatic alert. The video server can automatically send an e-mail with an alarm image to selected e-mail addresses, so the authorized people can take timely actions when they receive the information.

The integration potential of a digital surveillance system is another key benefit. This not only allows the surveillance system to share the network with other BA systems or the intranet network of the organization in order to reduce the installation cost of the surveillance system, but also provides possibilities for interoperation with other BA systems to achieve enhanced security, safety and advanced services for building users. For example, when a detector of the fire-detection system triggers an alarm, the CCTV system can be programmed to display images of the space associated with the fire detector so that the operator can very quickly find out if there is a real fire or if it is only a false alarm.

Remote accessibility is another main benefit. By connecting the cameras to a network, operators can see surveillance images from any computer on the network without the need and expense of additional hardware or software. If Internet connection is provided, you can be connected from anywhere in

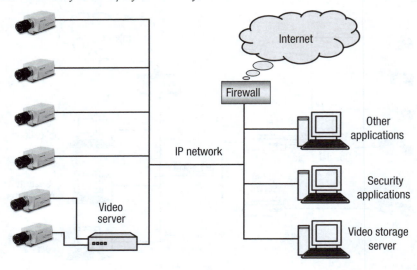

Figure 11.4 An IP surveillance system employing analogue and digital IP cameras.

the world to view a chosen facility or even a single camera from your surveil-lance system. Password-protected access can be set up allowing images and information to be kept secure and viewed only by authorized persons.

Two typical configurations of surveillance systems are commonly available today concerning the type of camera and its network device. Figure 11.3 shows the configuration of an IP surveillance system using analogue cameras connected to the IP network using video servers. The video servers provide the connection between the analogue cameras and the IP network. A video IP server may support one camera or multiple cameras. The video storage server, usually a computer, stores the video records, which are available for access by other applications on the network. Remote access is possible via Internet connection through a firewall. Figure 11.4 shows the configuration of an IP surveillance system employing both analogue and digital IP cameras. The IP cameras are connected to the IP network directly without the need for coaxial cables. In Figure 11.4, both digital IP cameras and analogue cameras supported by video servers are used in a single IP surveillance system.

When there are many cameras used in a system or the surveillance sys-tem shares the IP network with other BA systems or the intranet, network bandwidth is an issue. The bandwidth occupied by a live camera depends on many factors including the image-compression technology, the resolution of the image and the number of frames transmitted per second. For today's technology, it might be considered as a performance reference that a mini-mum bandwidth of 120 Kbps is required to support the transmission at a frame rate of 30 frames per second. Hard-disc storage requirements depend strongly on the frame rate of the video wanted to be stored.

11.2 Access-control systems

Access control is the ability to permit or deny the use of a particular resource by a particular entity. Physical access by a person is controlled on the basis of authorization, payment and the like. In physical security, the term access control refers to the practice of restricting entrance to a property, a building or room to authorized persons. Physical access control can be achieved through mechanical means such as locks and keys, or through technological means such as advanced access-control systems. Access control in this chapter is concerned only with physical access control.

Locks and keys have been used to control access to buildings and rooms for hundreds if not thousands of years. Today, the traditional key-based lock is still the most popular means used for access control of buildings, rooms and even commercial spaces. However, electric (or electronic) locks are commonly used nowadays to provide more effective or more secure access control. Electric locks are sometimes standalone with an electronic control panel mounted directly on the lock. More often electric locks are connected to an access-control system. Access-control locks are also designed in different forms to suit different applications such as to control access to public transportation, car parks, construction sites and even for immigration control.

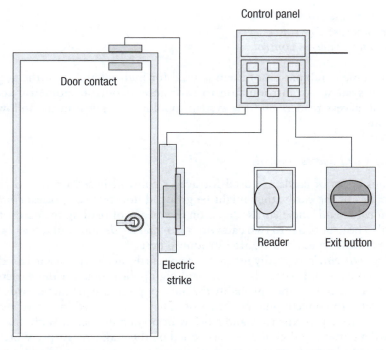

Figure 11.5 Basic components and configuration of a typical door access control system.

Electronic access control uses computers to overcome the limitations of mechanical locks and keys. A wide range of credentials can be used to replace mechanical keys. The electronic access-control system grants access based on the credential presented. When access is granted, the door is unlocked for a predetermined time and the transaction is recorded. When access is refused, the door remains locked and the attempted access is recorded by the access control system or even by a CCTV system. Figure 11.5 shows the basic components and configuration of a typical door access control system. When a credential is presented to a reader, the reader sends the credential's information, typically a series of numbers, to the control panel (a digital processor). The control panel compares the credential's number to an access-control list, and grants or denies the request. When access is denied based on the access-control list, the door remains locked. If there is a correct match between the presented credential and one in the access-control list, the control panel sends a signal to unlock the door.

There are usually three groups of credentials used for access control, referring to something a person knows (such as PIN, Personal Identification Number), something a person has (such as an access card), or a biometric feature (something the body of a person has, such as a fingerprint). Accordingly, access-control systems typically employ:

- *PIN access control*;
- *card access control*;
- *biometric access control*.

Often some combination of them is used for high reliability of the access-control system. An introduction to card access control, biometric access control, access readers and the system topologies is given in the following sections.

11.2.1 Card access control

Various types of cards are used for access control. Based on the working principles of the cards, they might be grouped into two categories: *conventional cards* and *smart cards*. Based on the means of reading the cards, they can also be grouped into two categories: *contact cards* and *contactless cards*. Some typical types of cards are introduced here.

Magnetic cards, typically magnetic stripe cards, are a type of card used for access control and particularly for banking applications. Of the variety of magnetic cards available, probably the most popular are magnetic cards that are similar to conventional credit cards. They can be used in a card reader that is relatively inexpensive and has few or no moving parts. With this type of card a pattern of digital data is encoded on the magnetic stripe. When the card is withdrawn from the reader, it moves across a magnetic head, similar to a tape recorder head, that 'reads' the data, sends the information to the system processor for verification, and if valid for entry at that point and that time, the processor sends a signal to release the door.

Wiegand cards, which look like credit cards, are widely used in access-control systems. They work according to a principle similar to that used in magnetic-stripe cards. Instead of a band of ferromagnetic material, the Wiegand card contains a set of embedded wires. The wires are made of a special alloy with magnetic properties that are difficult to duplicate. The set of wires can contain data such as the user's identification information, credit card numbers, medical history and so on. The card is read by passing it through, or bringing it near, a reader. The Wiegand effect occurs over a wide range of temperatures, so access-control devices using this technology can function in hostile environments. Other features include rapid response time and portability. These features make Wiegand cards and readers ideal for use in the field.

Proximity cards are another type of commonly used contactless cards using integrated circuits (IC) embedded in a high-grade fibreglass-epoxy card. The IC, capacitor and coil are connected in parallel. The card reader presents a field that excites the coil and charges the capacitor, which then energizes and powers the IC. The IC then transmits the card number via the coil to the card reader. The card readers communicate in Wiegand protocol. The earliest cards have 26 bits and the latest cards have an increased bit number to provide unique numbers.

Smart cards are a new type of card. Each card contains an integrated circuit chip embedded in the plastic of the card. This type of card has both a coded memory and a microprocessor with inherent intelligence. The card acts as a super-miniature computer as it records and stores information and personal identification codes in its memory. Smart cards can have a storage capacity of 8 KB up to 64 KB. The cards include contact smart cards and contactless smart cards. Contactless smart cards have a working range of 10 cm and are typically used for access-control applications. There are benefits to using smart cards for access control. Contactless smart cards can achieve a higher security level for the credential and the overall access-control systems. They also provide more storage and secure reading and writing of data. The capacity to add other applications to the cards, such as biometric identities, is one of the most important advantages of contactless smart cards over proximity cards. Radio-frequency identification (RFID) cards are a very common type of card nowadays. Users often are confused as to the difference between RFID and contactless smart cards. However, RFID, specified by a different standard, is usually used for low-security applications such as a supply chain. The other important difference is that the RFID has four options of working distance ranging from 10 cm to 10 m (corresponding to four working frequencies of range between 125–135 kHz and 2.45 GHz).

11.2.2 Biometric access control

Each person has unique biometric features which can be used for personal identification. The typically used biometric features include hand geometry, fingerprints, palm prints, voiceprints, face recognition, signature verification and iris recognition. Biometric access control using well-developed biometric identification technologies can provide a high level of security for access control. There has been much work carried out on developing biometric identification technologies. The key issues of a biometric identification technology are the computational efficiency and reliability. Today's technologies can provide biometric identification of quite high reliability and computational efficiency, and have therefore attracted wide applications in the area of access control in security.

11.2.3 Intelligent readers and system topologies

An access-control system of any type needs readers as the essential elements of PIN entries, to read information from cards, or to collect biometric information. According to their means of integrating the readers with the access controllers and the roles of the readers, readers may be classified as *conventional readers*, *semi-intelligent readers* and *intelligent readers*.

Conventional readers (or basic readers) have no built-in intelligent functions. They simply collect the information and forward it to a control panel. Wiegand protocol is commonly used for transmitting data to the control panel. Other examples of common options include RS-232 and RS-485. Figure 11.5 shows an example of an access-control system using a conventional reader where separate connections to the control panel are provided for the door contact, lock control and exit button.

Semi-intelligent readers, besides reading the information, provide all other inputs and outputs (including lock control, door contact and exit button) necessary to control door hardware, but do not make any access decisions. When a user presents a card, PIN or biometric identity, the reader sends information to the main controller and waits for its response. The protocol for connecting semi-intelligent readers and control panels is typically RS-485.

Intelligent readers provide the input and outputs and have the same means of connecting to control panels as semi-intelligent readers. The main difference is that they have information-processing ability to make access decisions independently. They are also typically connected to a control panel via an RS-485 bus. The control panel sends configuration updates and monitors events from the readers. Figure 11.6 shows an example of an access-control system using a semi-intelligent or intelligent reader where the door contact, lock control and exit button are connected to the local reader. In the case of using a semi-intelligent reader, the access decision is made by the control panel, while the decision is made in the local reader in the case of using an intelligent reader.

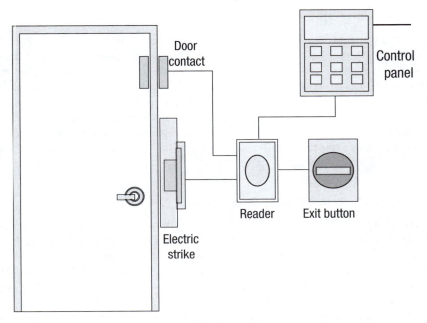

Figure 11.6 Typical configuration of access control systems using semi-intelligent readers or intelligent readers.

In access-control systems in practice, one control panel may be connected to multiple sets of readers and inputs/outputs or to multiple intelligent readers for access control of multiple spaces. For medium and large access-control systems, many control panels will be used. In these cases, different options of different network configuration may be selected for the systems of different manufacturers, providing different communication efficiency and reliability. The issues concerning control network configuration are discussed in Chapters 4 and 6 related to local area network. RS-485 employing different topologies is typically used for access control systems, while IP controllers and IP readers integrated and connected to a central PC station via IP LAN or Internet are becoming available in the market.

Wiegand protocol is a protocol commonly used for communication between non-intelligent readers and their control panels based on a Wiegand interface. The Wiegand interface uses three wires. One is a common ground and the other two are data-transmission wires usually named as Data High and Data Low. The high voltage level is usually +5 V DC accommodating long cable connections from the readers to their control panel (maximum length up to 150 m). An advantage of the Wiegand signalling format is that it allows very long cable connections compared with other interface standards for the same applications.

11.3 Burglar alarm systems

The use of burglar alarm systems (or intrusion-detection systems) is to detect unwanted attempts in accessing a space or object. The main functions of burglar alarm systems can be divided into three categories including: *perimeter protection, area/space protection* and *object/spot protection*. Various sensing devices of very different mechanisms are available for detection at different situations. Intrusion-detection systems also often refer to the systems for protecting computers or other information systems from unwanted access. However, this chapter is only concerned with the detection of unwanted physical access to protected spaces or objects.

11.3.1 Functions of burglar alarm systems

Perimeter protection is usually achieved by mounting sensing devices on doors, windows, vents, skylights or any openings to a business or home. The advantage of perimeter protection is simplicity of design. Its disadvantage is that only the openings are protected. Typical sensors used for perimeter protection include door contact, electric field fence sensor, infrared beam sensor and glass break detectors.

Area/space protection systems are designed to protect the interior spaces of a facility. Sensing devices used for space protection are particularly effective against 'stay-behind' intruders. The detection sensors can be classified into four main categories including audio, pressure, electronic vibration and motion detection. The advantage of area/space protection systems is that they provide a highly sensitive, invisible means of detection. Their disadvantage is that incorrect application and installation can result in frequent false alarms. In many practical applications, space protection is used as a backup to the perimeter protection system. Typical sensors used for area/space protection include passive infrared detector (PIR), photoelectric beams, ultrasonic detector and pressure-sensitive mats.

Object/spot protection aims at providing direct protection for specific items, including direct security for high-value items. Such a detection method is the final stage of a comprehensive protection system. Typical sensors used for object/spot protection include capacitance/proximity detectors and electronic vibration detectors.

11.3.2 Alarm control panel and alarm control

All sensing devices are wired into the alarm control panel, which receives their signals and processes them. The type of control panel needed is dependent upon the sophistication of the overall intrusion alarm system. Some control panels provide zoning capabilities for separate annunciation of the sensing devices. They may also provide low-voltage electrical power for the sensing devices.

A modern control panel uses one or more microprocessors. This allows the control panel to send and receive digital information to and from the alarm stations. Each authorized user can also have his or her own unique code or present his or her identity to activate or deactivate the system. If the system is not deactivated within a pre-set time after the system is triggered, the system will generate alarms. Local noise or visual alarms are usually generated. In the meantime, alarm signals will be sent to the monitoring centre or delegated remote security mangers via LAN, telephone, Internet or other means.

11.4 Fire alarm systems

The function of fire alarm systems is to detect the presence of unwanted fire in the protected spaces by monitoring environmental changes associated with combustion. Fire alarm systems may be activated automatically, manually or usually both. The purpose of using fire alarm systems is to notify people to evacuate in the event of a fire or other emergency, to call the fire protection department for emergency aid, and to activate other associated systems to control the spread of fire and smoke. It is worth noting that the fire alarm system is an essential measure but not the only measure for the fire safety of a building in terms of both regulations and reality. Concerning fire alarm systems, it is critical to properly select and place the detectors according to the layout and use of spaces. This section discusses the basic components and the typical configurations of fire alarm systems.

11.4.1 Typical fire detectors

A fire alarm can be initiated manually or automatically. Manually operated devices, such as a 'break-glass' alarm, provide the means for occupants to activate the fire alarm system when they observe a fire or smoke. Automatic fire detectors commonly used can be summarized into the following types, including: *heat detector (or heat-sensing fire detector), smoke detector (or smoke-sensing fire detector)* and *flame detector (or flame-sensing radiant energy fire detector)*. Different types of detectors have different detection speeds and probability of false alarms. Different types may be used to increase the detection speed and to enhance system reliability.

Heat detectors: A heat detector detects fire by sensing changes in ambient temperature. Typically, if the ambient temperature rises above a predetermined threshold an alarm signal is triggered. Heat detectors may work on the basis of rate-of-rise of temperature, fixed temperature or both. *Rate-of-rise heat detectors* are activated by the sudden rise in ambient temperature. A sudden temperature rise above a change-rate threshold, such as 8 K per minute, will activate the alarm. *Fixed-temperature heat detectors* are activated when the ambient temperature reaches a fixed threshold, such as 58°C. Heat detectors are usually installed in spaces, such as kitchens or utility areas, laundry rooms or garages, where smoke and fire detectors should not be installed. This will allow extra time to evacuate the building or to put out the fire if possible.

Smoke detectors: Smoke detectors detect smoke and issue signals to fire alarm systems. There are many different smoke detectors based on different mechanisms and designs. Common types include: *ionization smoke detectors, photoelectric (optical) smoke detectors, air-sampling smoke detectors* and *carbon monoxide/carbon dioxide detectors.*

An *ionization detector* contains a small amount of radioactive material that ionizes the air between a positive and negative electrode. The conductance between the electrodes is measured. Introduction of smoke into the sampling chamber of the detector reduces the conductance between the electrodes. When the conductance falls below a pre-set threshold, the detector is triggered.

A key element of a *photoelectric detector* is the light sensor. In a smoke detector, the light sensor is provided with a light source (e.g. infrared LED) and a lens to concentrate the light into a beam at an angle to the light sensor. In the absence of smoke, the light passes in front of the detector in a straight line. When smoke enters the optical chamber across the path of the light beam, some light is scattered by the smoke particles and directed at the sensor; thus the alarm is triggered.

An *air-sampling smoke detector* detects microscopic particles of smoke. Most air-sampling detectors are aspirating smoke detectors, which work by actively drawing air through a network of small-bore pipes laid out above or below a ceiling in parallel runs covering a protected area. Small holes drilled into each pipe form a matrix of holes (sampling points), providing an even distribution across the pipe network. Air samples are drawn past a sensitive optical element to detect the extremely small particles of combustion. These types of sensors are used in high-value or mission-critical areas. Smoke detection systems using air-sampling smoke detectors can provide multiple levels of alarm threshold, such as alert and action, and therefore achieve high sensitivity in smoke detection.

Flame detectors: Flame detectors detect flames directly by using optical sensors. Common types of flame detector include: *ultraviolet (UV) flame detectors, infrared (IR) flame detectors* and *UV/IR flame detectors. Ultraviolet flame detectors* work with wavelengths shorter than 300 nm. These detectors detect fires and explosions within 3–4 ms due to the UV radiation emitted at the instant of their ignition. False alarms can be triggered by UV sources such as lightning, arc welding and sunlight. In order to reduce false alarms a time delay is often included in the detector design. *Infrared flame detectors* work within the infrared spectral band. Hot gases emit a specific spectral pattern in the infrared region, which can be sensed with a thermal imaging camera. A *UV/IR flame detector* is the combination of ultraviolet detection and infrared detection, which confirms the fire by using UV and IR thresholds in 'AND' configuration to minimize false alarms. A comparison of the characteristics of typical automatic fire detectors is shown in Table 11.1.

Table 11.1 A comparison of typical automatic fire detectors

Type of detector	Detection speed	Probability of false alarms	Detector cost
Heat	Slow	Low	Low
Smoke	Fast	Medium	Medium
Flame	Very fast	High	High
Particle sampling	Fast	Low/Medium	Medium/High

11.4.2 Fire alarm control panels and their topologies

A *fire alarm control panel*, or fire alarm control unit, is an electric, nowadays typically computer-based, panel that is the controlling centre of a fire alarm system. The panel receives information from automatic and manual fire detectors, monitors their operational status and provides automatic control and transmission of information necessary for fire suppression and alarms based on a predetermined sequence. The panel may also supply electrical power to operate associated sensors, transmitters and activation devices.

Typical types of fire alarm control panels used include: *coded fire panels*, *conventional fire panels*, *multiplex fire panels*, and *addressable fire panels*. *Coded fire panels* were made prior to the 1970s, and are rarely used nowadays. They have been largely replaced by modern conventional fire panels.

Multiplex fire panels are something of a transition between conventional and modern LAN-based addressable fire systems. They have usually been used for large or complex buildings. Multiplex fire panels, generally replaced by modern addressable fire systems, are not built for new installations any more but they are still in use in many existing buildings. Similar to modern addressable fire systems, they can address individual sensing and control devices, but in a less efficient manner.

11.4.3 Conventional fire panels

Conventional fire panels are rarely used in large buildings nowadays, but are still commonly used in smaller buildings such as small schools or apartment buildings. A conventional panel usually consists of a few loops, each involving a few sensing or actuation devices. Each loop is responsible for a zone within a building. The main drawback of conventional fire panels is that they can identify an entire activated loop but cannot identify which sensing device has been activated within a loop.

Conventional fire panels use analogue electrical signals with a few common configurations to integrate the sensing devices. Figure 11.7 shows the connection method of a conventional fire panel which connects the sensing devices in series in closed loops. Figure 11.8 shows the connection method of a conventional fire panel which connects the sensing devices in parallel in

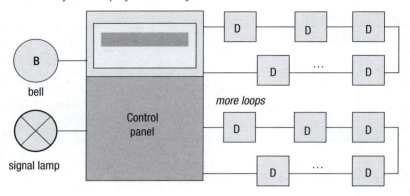

Figure 11.7 A conventional fire panel connecting sensing devices in closed loops (D: fire detector).

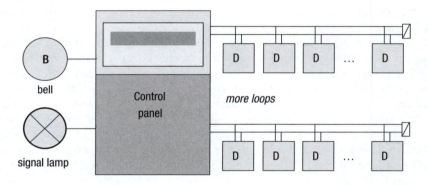

Figure 11.8 A conventional fire panel connecting sensing devices in open loops (D: fire detector).

open loops. Fire detectors (smoke detector, heat detector, gas detector etc.) and actuation devices (e.g. fire bell, guide light, sprinkler, smoke ventilator etc.) are connected to the control panel via a dedicated analogue signal line with a current of 4–20 mA. In particular, several fire detectors that guard the same area are connected to a single signal line. If one sensing device in a loop senses a fire and is activated, the panel will note the fire in that zone by detecting the abnormal electrical characteristics of the loop. But the fire alarm panel cannot distinguish which detector in that zone triggered the alarm.

Figure 11.9 shows the connection method of a conventional fire alarm panel based on an initial device circuit (IDC, Class B) defined by NFPA 72 (2007) of the National Fire Protection Association (NFPA) in the United States. As shown in the figure, the fire panel perceives a fire based on the fact that the current increases in the corresponding analogue connection line when a fire occurs. In this case, the fire detector that perceives the fire causes the analogue line to short-circuit so that the current increases. At this

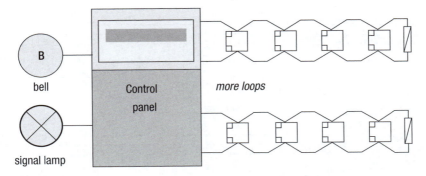

Figure 11.9 A conventional fire alarm panel based on an IDC (Class B).

time, when the fire panel detects that the current in a line has increased, it relays this information using a bell or a signal lamp to indicate that a fire has occurred in the corresponding area. There is also a definition of Class A configuration providing higher reliability but with two more cable lines which can be relatively more costly.

11.4.4 Addressable fire panels

Addressable fire panels, or network-based fire panels, use wired or wireless LANs to connect the sensing devices and actuation devices to the panels. Each of the automatic and manual fire detectors as well as the actuation devices, as processor-based smart devices, has a unique address in the network, allowing the fire control panel to identify which detector has triggered a fire alarm and which detector has a failure. Typically, two-wire network connection is used. Four-wire and three-wire networks may be used for the detector loops, but they are not common. Addressable fire panels are usually used nowadays for large building systems.

Bus topology and ring topology are usually used for the sensor loops. Figure 11.10 shows a fire alarm panel using bus topology to connect the detectors and actuation devices. Figure 11.11 shows a fire alarm panel using ring topology. The fire alarm panel and its smart detectors and actuation devices communicate in a master–slave mode. The fire panel, as the master, supervises the communication and responds to significant state changes with appropriate actions. As an example, the German standard, DIN EN 54, requests no more than 32 detectors in a detector loop to guarantee that no more than 32 fire detectors are affected by a single short circuit or cable break. One single fire alarm panel may involve up to 20 or more detector loops, but typically less than ten loops. The scanning frequency of the fire panel on the individual detectors associated is an important factor to ensure the timely response and reliability of the fire alarm system. The above standard requests a minimum scanning frequency of once in 10 seconds. The

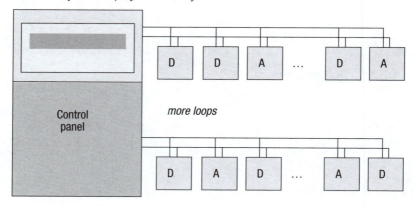

Figure 11.10 An addressable fire alarm panel using bus topology at detector level (D: fire detector, A: actuation device).

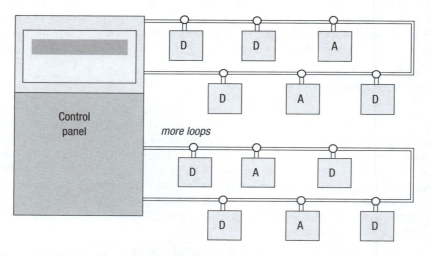

Figure 11.11 An addressable fire alarm panel using ring topology at detector level (D: fire detector, A: actuation device).

scanning frequency of systems in practice, however, is usually much higher than this minimum requirement.

Large buildings may need to use a large number of fire alarm panels to be integrated via a higher-level automation network. A single KNX/EIB system allows a maximum 225 detector loops at the field level. A single domain of LonTalk allows a maximum 255 detector loops at the field level. These are equivalent to a maximum of 7,200 and 8,160 detectors or actuation devices, respectively, sufficient even for very large building installations.

Traditional addressable fire alarm systems use proprietary protocols. There have been significant efforts made in developing and adopting standard open protocols and/or extending the standard building automation protocols to suit the application of fire alarm systems at both detector network level (field network level) and alarm panel network level. Current open network protocol standards addressing the applications of fire alarm systems include BACnet, LonWorks and KNX/EIB. More work is needed on enhancing the standards themselves and more promotional efforts are required on adopting open standard protocols commonly in fire alarm systems in the market. However, it is probable that the day will come soon for fire alarm systems when open standard network protocols are commonly used by the products in the market, like today's building environmental control systems.

11.5 System integration and convergence

Modern physical security systems may comprise access control, CCTV, fire detection and burglar alarm systems, preferably integrated into a single solution. Traditionally, these systems were dominated by proprietary hardware and software solutions. Nowadays, open and standard solutions are the trend and what the market demands. Software and system architecture are what differentiate modern systems from older ones.

Thanks to developments in software engineering, most of the application software for security systems is now very user-friendly and offers high performance and capability, flexibility and compatibility with other applications. The cost of software development has reduced, resulting in more attractive prices to end-users. The cycle for upgrade and enhancement is much shorter. New features and capabilities can be added and delivered to end-users more quickly.

As discussed in Chapter 5, integration and interoperation of different subsystems from the same manufacturer or different manufacturers can be achieved at control network level (field network level) and at management level with simple physical connection for very simple interoperation. The integration at management level is to build the communication between software applications on (typically different) computers used as the central management stations of the subsystems. As there are standard and effective software interface technologies (see Section 5.8) available in the market, which are reliable and convenient to use, most modern management software applications have adopted some of these communication interfaces. This means that different security control subsystems and other BA subsystems can be integrated at management level with less difficulty. The popular use of IP networks provides great ease of integration as well.

There has been good practice in using standard technologies for CCTV systems, particularly for image processing, even at the stage of analogue systems (such as VHS and DVD). After recent implementation of IP-based digital CCTV systems, compatibility between systems from different suppliers is much more feasible.

However, integration of other security control systems at control network level has been a very different story. Integration at control network level requires the use of the same network protocols in all layers. For systems from different manufacturers, integration at this level requires in practice the same open and standard protocols to be used. The progress of adopting open and standard protocols in the security control market lags far behind the environment control industries (e.g. HVAC control and lighting control). The security market has not adopted many standards for communication between systems of different manufacturers. The primary efforts on standard protocols in security systems have been at the card and reader level (such as Wiegand) and the biometric identification level, but not at the network or system level.

When BACnet and LonMark were developed, security applications were not initially considered, and protecting the data was not part of the design. Safety systems have felt comfortable using BACnet, since BACnet can present critical data as 'read-only'. The fire panels make the critical decisions on their own without the need for being reprogrammed on a regular basis. In an access control system, decisions and programming are constantly being made from a workstation that needs to securely communicate the instructions and updates to the field panels. An access control system is programmed whenever a card record or a door schedule is changed. Information about who enters or leaves a facility is highly sensitive and should be available to only the appropriate operators. There have, however, been effective moves to adopt open and standard protocols at this level, such as the efforts of the BACnet committee of ASHRAE and the Life Safety and Security Working Group (WG-LSS).

References

Axis Communications. (2002) *Converting an Analog CCTV System to IP-Surveillance*, Axis White Paper, Lund, Sweden: Axis Communications AB.

Bernard, R. (2004) The convergence of physical security and IT: broad convergence – IT, security and building controls, *Security Technology & Design Magazine*. Available at: www.go-rbcs.com/Convergence7.htm

Fennelly, L. J. (2004) *Effective Physical Security*, 3rd edn, Boston, Massachusetts: Elsevier Butterworth-Heinemann.

Gagnon, R. M. (2008) *Design of Special Hazard and Fire Alarm Systems*, 2nd edn, Florence, Kentucky: Thomson Delmar Learning.

Lee, K. C. and Lee, H. H. (2004) 'Network-based fire-detection system via controller area network for smart home automation', *IEEE Transactions on Consumer Electronics*, 50(4): 1093–1100.

Neugschwandtner G., Kastner W. and Erb, B. (2006) 'Fire safety alarm transmission in networked building automation systems', 2006 IEEE International Workshop on Factory Communication Systems, pp. 79–82.

Norman, T. (2007) *Integrated Security Systems Design*, Boston: Elsevier Butterworth-Heinemann.

Wikipedia. (2009) 'Access control'. Available at: http://en.wikipedia.org/wiki/Access_control

Wikipedia. (2009) 'Burglar alarm'. Available at: http://en.wikipedia.org/wiki/Burglar_alarm

Index